LIFE AND DEATH ON MARS

THE NEW MARS SYNTHESIS

John Brandenburg Ph.D.

Other Books of Interest:

THE CASE FOR THE FACE
EXTRATERRESTRIAL ARCHEOLOGY
INVISIBLE RESIDENTS
THE COSMIC WAR

LIFE AND DEATH ON MARS

John Brandenburg Ph.D.

Adventures Unlimited Press

LIFE AND DEATH ON MARS

by John Brandenburg Ph.D.

ISBN 13: 978-1-935487-36-4

Published by:
Adventures Unlimited Press
One Adventure Place
Kempton, Illinois 60946 USA
auphq@frontiernet.net

www.adventuresunlimitedpress.com

Table of Contents

Picture credits: All pictures are taken from Wikipedia Commons unless otherwise marked

Dedication :

To the generations of humanity who will be born and grow up on other worlds

Foreword: Oasis Earth

"One should not increase, beyond what is necessary, the number of entities required to explain anything."

Bishop William of Ockham (c.1285–c.1349)

The planet Mars (JPL/NASA)

Bishop William of Ockham

Given our present understanding of the cosmos, it can be predicted that in some future time, based on sciences not known to us at present, something resembling the following will be published one day in a journal of archeology:

> *I am writing quickly now. In the year of 3 reeds, in the month of 7 flowers, on the day of bells, did the evil star appear in the constellation of the Deer [our Orion] it grew brighter in the nights following, and moved into the constellation of the Spider [our Taurus] there it rested and grew brighter still, near the Ruler of the Night [believed to be Sirius]. The chief priest observed the star for several nights and said that it was like the evil star that was seen before the great*

dying time in the fourth dynasty and the time of no records. On the seventh day of our seeing, the oldest of the star gazers was brought to the temple and observed the star and its tracing amid the stars, by the order of the Emperor, and he said also this star was like unto the killing star seen before the great dying, only it was more mighty. He then went to the Emperor and told him "this is the end." The Emperor was angry at this report and ordered the man slain.

It is now twelve days since the sighting of the great star, which because of its brightness and movement across the sky in the manner of a killing star, has caused much fear and despair. It is this night or a night soon to occur that the killing star will fall, as said the stargazers, because of its track across the stars. It has been three days since the Emperor ordered that all the wise men and magicians of the court be put to death. For, he said, their spells and wisdom are useless to stop this killing star. The star is now seen even by the common people and they are much afraid. Many sacrifices have been made in the temples, the most beautiful of the young offering themselves willingly, yet the star has not departed. So it is now said this night it will likely fall from the sky.

Now comes the star and becomes very bright, blinding, ...[final word fragment- possibly thunder]

(End of record)

From the last entries of the Codex Rosetta recovered from the great Pyramid. Text Translation by Tohsiro Tanaka using the Braginski and Lewis translation method.

So may read our descendants one day of the last record of an extraterrestrial civilization, whom the cosmos has first nurtured then destroyed.

Only one meteorite is reported to have killed a living thing, and it came from Mars.

In the oasis of Nakhla Egypt, far out in the desert, away from the Nile where the Sphinx stands guard over the Pyramids, a stone fell one morning in 1928. It killed a dog. Fifty years later, it would be discovered that this meteorite had another remarkable distinction besides being a bringer of death: Of the millions of known meteorites, it belonged to a small group of one dozen that had their origin on Mars. This remarkable coincidence tells the pure scientist almost nothing, but to the poet it speaks volumes. This book is about the deciphering of that message Mars sent us in a rock, a story of life and death, and this book will also depict the journey to that knowledge through the ages.

Mars (Ares) the God of War

Two concepts, so coherent and important down the paths of time that they are commonly depicted as persons, will be our traveling companions and guides on this journey. The first is the personage of Mars, the god of war and all that pertains to it, typically personified as masculine. He is a dread and dark personage, but one must listen to his instruction and not tremble, for he has no patience with cowards.

The second personage to accompany us is Wisdom, a feminine personage in the ancient texts, including the Bible. Whether as Athena, the patron goddess of Athens, the seat of ancient learning and reason, or as Sophia, the spirit of wisdom in the Proverbs, she and those who loved her will brighten our way. She is a gentle soul, mostly, who loves

Athena, Goddess of Wisdom

all that lives, and rejoices in calm, reasoned, discourse—so different from Mars, yet like him in her love of useful truth. This is a journey of knowledge, guided by wisdom, concerning new paths and frontiers for humanity, but the wisdom and knowledge we seek is not abstract and academic. Mars is giving us knowledge on matters of life and death.

This book is about the power of knowledge to preserve life: our own. It is a book about finding truth, not only scientific truth, but poetic truth as well, truth that guides one's life, not just one's path through space. Scientific truth can be tested in a bottle. Poetic truth must be lived. Poetic truth is holistic truth, truth that looks at the whole picture formed by a million pixels, while science tends to dwell on the pixels themselves, individually. Both kinds of truth are essential, we shall discover. For the cycle of hypothesis and its testing that science uses is required to establish the true pixels that make up an image. This ruthless discipline in the search for truth stems from the scientist's companionship with Mars, whose battlefield was the ultimate testing place of all hypotheses. If your science is bad, your people will drink of the bitter dregs of defeat. This association has led to science being, on occasion, a full contact sport, where the search for truth has been submerged in a battle of personalities and schools. However, occasionally science tries to look at the whole picture, to grasp a view of the forest, rather than merely studying the individual trees. On these occasions, the scientist seeks the companionship of Wisdom more than any other, whether she speaks through dreams or lines on a graph. At these points of trying to grasp the whole picture, the scientist and that other searcher for truth, the poet, find the most in common, and the scientist will speak in terms of "elegance" of a theory and the poet will speak of "mathematical beauty."

When a scientist tries to fit a mass of scientific data, gathered from many different sources, into a whole picture, this is no longer called a hypothesis, it is called a synthesis. A synthesis is not hard edged and detailed like a hypothesis; it is by nature a story, a broad vista. It is a forest or woods rather than a collection of trees, sprawling over the plains and mountainsides. A synthesis is by nature indistinct around its edges and shadowy in its rich depths. Science labors continually under the influence of syntheses, new and old, sometimes barely articulated.

Fragment of the Nakhla Meteorite (NASA)

Ancient Alexandria, Egypt, at the Mouth of the Nile

A key tool of the scientist in finding truth is Ockam's razor, named for Bishop William of Ockam, who helped shape the scientific revolution. His idea of a "minimum of entities" is often translated as the admonition that one should "choose the simplest hypothesis." This is reasonable interpretation, and very useful; however, it loses the point that truth is not simple. Many things that are commonplace, and whose existence and action underpins reality, are complex and marvelous. The study of the cosmos around us has revealed that even things that look simple and provide the bricks with which we build our understanding of things—an atom, a star, or a person—are not simple. William of Ockham knew this. For this reason he spoke of a minimum of "entities" being required to explain what is seen. Entity is a word well chosen by the Bishop, for an entity is where the complexity upon which reality is based can be loaded. An atom is an entity, so is lightning, a galaxy, and a person. All of them are complex, not simple. The tension between these two translations of Ockam's razor, that of his original words, and its modern everyday paraphrase, forms the basis of the scientific conflict we will encounter on this journey of discovery.

We will find as we explore Mars that we now encounter a crisis of science. This crisis has its origin in the disagreement over how to apply Ockham's razor. This disagreement centers on the definition of the word "entity." To some, "entity" means only chemical elements and their reactions, to others entity has a general meaning of any phenomenon that can be isolated and characterized. It is the question over whether biology is an entity, or are the chemicals which make up its being. Is biology the simplest answer to a question, because it is one entity, or is it a cosmos of innumerable and improbable things and processes and, therefore, too complex to be an answer. Over this question has emerged a deadlock in science. To break this deadlock a New Mars Synthesis has been formed.

The formation of a new synthesis, from the Greek words "syn"—together—and "thesis"—idea—is a rare occurrence in science, coming only after a critical mass of new knowledge is available. It is the assemblage of the mass of new facts into a whole story. On Mars, a planet that has mesmerized humanity since ancient times, our scientific investigations have arrived at such a point where a new synthesis can be proposed. This book is a presentation of this synthesis, how it emerged, and its meaning. It is a deciphering of a message that arrived in a stone flung from Mars that fell in Egypt and killed a dog. The message is a lesson for us. Mars's first words of this lesson are *"in the cosmos are the quick and the dead. Do not dwell carelessly, my children ... be quick."*

Life and Death on Mars

Chapter 1. The School of Mars

"Because of our sins, an unknown nation came against us, from an unknown land ."

Russian Account of the Mongol Conquest

"And he made in Jerusalem engines, invented by cunning men, to be on the towers and bulwarks, to shoot arrows and great stones…"

II Chronicles 26: 14, The Bible

In ancient Egypt the war god was named Anhur, who was also god of the sky, later subsumed into the falcon-headed god Horus, as the kingdoms of Egypt merged. Horus was called Horus the Red and was identified with the planet Mars, considered the chief of the "stars that know no rest," as the planets were known to the Egyptians. Horus was also identified with the metal iron. Iron was called "the sky metal" in ancient Egypt and in many other ancient cultures because it was first known from meteorites, and only later smelted from terrestrial ore.

Horus, the war god of Egypt.

Mars has been both a bright red light in the sky and a human concept of living since ancient times. Humans are, at their worst, social predators, like the lion, the wolf, and the ant. Mars was the personification of this social predatory instinct expressed against fellow humans, who are the most dangerous creatures on the planet. First as Nirgal in Mesopotamia, then as Ares, finally as Mars, in the Mediterranean basin, the planet Mars was the heavenly light associated with the god of war. It owed this honor to

Ancient weapon of mass destruction

its blood red color and its close proximity in orbit to Earth, which meant that when it was in opposition, its closest point to Earth (see figure below) it would become the brightest object in the night sky. To the Romans he was represented by the male symbol, which, by some accounts, represented his shield and lance.

The symbol of Mars

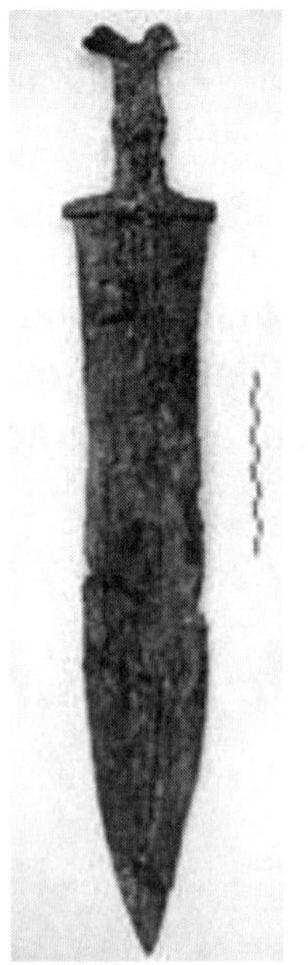

Ancient Roman sword

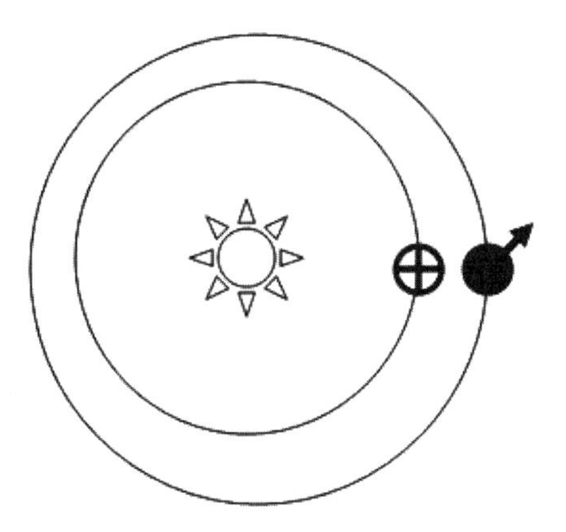

The orbits of Mars and Earth around the Sun, with Mars at opposition.

Mars, the Roman god of war

Mars is the next planet out from Earth in the solar system, followed by the asteroid belt, and then Jupiter, which is brilliant but an unremarkable yellowish-white like Venus. Venus, the next planet in, is bound, by being closer to the Sun, to be only an evening or morning star. Mars in contrast seems to roam the heavens freely and moves from one end of the night sky to the other. Every 2 years and 7 weeks Mars becomes dazzlingly bright. As Venus, the brilliant planet, evoked thoughts of love to the ancients, with its pale golden color leading to her being often pictured as a blonde, the red color of Mars evoked the sight of blood.

It is told that Ares (the Greek analog of Mars), the god of ultimate maleness, had a torrid affair with Aphrodite, the goddess of love. It was really a love triangle since Aphrodite was the wife of Hephaestus, at the time. He was god of volcanoes (called Vulcan by the Romans) and metal working.

To the Greeks, known for their science and philosophy, the war god Ares and the activities he represented were also very important. The Greeks were militarily very aggressive and inventive. Homer, the oldest literary figure of the Greeks, is not famous for a tale of peace, but for the story of Helen of Troy. The Fall of Troy was accomplished through the stratagem of the Trojan horse, and this tale may represent an embellished report of the use of a siege tower against a walled city.

The tale of the Odyssey celebrates the great accomplishments of the Greeks as the first great mariners of the Mediterranean. In that tale Odysseus, who was the favorite of Athena, the goddess of wisdom, journeys by sea on a great voyage of peril as well as discovery of strange new lands. His unfortunate journey, and the general misfortune of the Greeks returning from Troy, was due to the wrath of Athena, for the Greeks, in their destruction of Troy, had violated her temple by seizing Cassandra the prophetess there after she had sought sanctuary.

War, unlike nature, presents humans with a set of problems that are constantly and rapidly evolving. Since ancient times, war, or the threat of war has been a constant driver of technology and sciences. It has also been a source of much of humanity's misery, a fact not lost on the Greeks, who pictured Ares as a rascal and trouble maker.

To the Romans, who conquered the Greeks and took their mythology along with everything else of value back to Rome, Ares merged with an old Roman deity Mars supplanting that god's previous role as a god of agriculture. This transformation mirrored the elevated role of war in Roman society, as it transformed itself from a republican city-state mostly interested in farming into a sprawling empire guarded by a massive military. To the Romans, Mars represented the epitome of manhood, courage, boldness, cunning, discipline. The love triangle of Ares, Aphrodite, and Hephaestus became Mars, Venus, and Vulcan, and went from being a tale of scandal on Olympus to a cautionary tale for men who considered avoiding military service. The warrior gets the girl, and the honest workman is left with a new set of horns. To serve Mars was to become a human predator, and like all predators he cultivated a cunning and aggressive intelligence.

When war was declared in ancient Rome, a frequent occurrence, the doors of the temple of Mars were thrown open. The lance of Mars was then brought out before the crowds in a ritual display. In the adulation of the crowd was the decision of war validated.

Mars was a god who constantly inspired the search for new technologies. The sword, which we regard as the traditional symbol of things military, was in fact viewed as a remarkable innovation and sign of capable metal-working technology by the ancients. It was literally a "long knife" in the terminology of the Native Americans of the Plains and took lots of metal, either bronze or steel, and the ability to shape it into large pieces. This required a skill level and capability to obtain amounts of strong metal far beyond that required for hunting, fishing or protecting flocks from predators. For a wolf or lion, a big spearhead or arrow head, or even a sling stone, was all that was required to kill or drive them off. However, to fight the people in the next river valley, you needed swords. Woe unto the people whose skill at the creation of swords was inferior to its neighbors. As agriculture made the supporting of large urban populations possible, the walled city became the norm and to serve the god Mars properly one had to learn to defeat them. This led to the design and construction of siege engines, the first weapons of mass destruction to be developed.

To throw stones from a catapult against city walls effectively one needed to be able to predict ranges. To predict ranges for rocks one needed to know their mass. The Greeks discovered that a rock made round like the Sun or Moon had a mass that was proportional to the cube of its radius, and hence it has been found that the Greeks solved this problem by a graphic method for their catapult operators. The graph gave the mass of the spherical rock based on its radius. Thus, one of the first applications of Greek mathematics and geometry was in war.

The ability to build a strong wall to guard your city was a highly desirable skill in ancient times. The best walls were of large tough stones shaped into bricks of the same size and shape. Smaller or larger bricks were weak links in the chain of defense.

Such were the walls of Troy which the Greeks could not breach in ten years of war. Make your walls of weak stone, and come the time "when kings go forth to war," your neighbors will come with battering rams and break down your walls. So the school of Mars offered degrees in stonemasonry in ancient times as well as metallurgy.

To make a good wall the number of equal-sized bricks of good stone required mathematics. In particular this required that the diagonal measure of each section of wall be a whole number of bricks as well as the height and width. The Egyptians knew this when they constructed the pyramids, for the length of the faces of pyramids is a factor of five to three times the length of half the pyramid base, and likewise the face length is a ratio of five to four to its height. The half length of the base squared and height squared are the length of the face squared in a 3:4:5 ratio, or 9:16:25 when squared.

To the base of this pyramid came the philosopher Pythagoras and studying from it he generalized that the diagonal of any rectangle squared was the sum of the sides squared. This enabled stonemasons to build stronger walls for cities because they could get the number of bricks right. Pythagoras had a more than academic interest in helping the Egyptians build better walls at that time. The Persians invaded Egypt and conquered it while he was there and he was imprisoned. When released, he returned to Greece with tales of the cruelty of the Persians, and thus prepared Greece for its own war with the Persians, a generation later.

Pythagoras

His Inspiration

Later the Vikings developed their own written record of their religion, and Tyr, the god of war, was among their pantheon. He was a noble and fearless character, who was the only god of Asgard who could deal with the great wolf Fennir. When the wolf became too dangerous to run free, Tyr gave his right hand as pledge to the great wolf, placing it the wolf's mouth, while it was chained by trickery. When the wolf realized it was now chained, Tyr's hand was his sacrifice. In the last battle between the gods and the forces of evil at Ragnarok, Tyr dies valiantly while slaying the great hell-hound Garm.

Tyr the Viking god of war

Tyr was notable also as the god of the Allthing, the first parliament. It was in the Allthing that all things could be discussed freely and votes taken. This form of government spread from the Vikings to ancient Britain and became the basis for modern democracy. Tyr, under the name Tue, is honored now in the second day of the week, Tuesday, which, in the Latin tradition, is called Mars. His symbol dispensed with the shield and got straight to the point.

Tyr and the dog

The straight-to-point symbol of Tyr.

Tyr and his troublesome dog

The school of Mars did not end in ancient times. Even if Mars ceased to represent a god and instead, with the advent of a Christian Europe, became just a planet, a moving light among the fixed stars, it lost none of its old meaning. Mars's spell lived on in the astrological tables, especially in the astrologers' tables that gave advice to Kings.

The periodic oppositions of the planet Mars, 2 years and 7 weeks apart when it is closest to Earth and thus dominates the night sky, became occasions for the time "that Kings go forth to war," entailing the launching of wars of aggression for kings and nations who felt a need to invade their neighbors periodically. They would ask themselves, "if the sky itself appears to be provoking bloodlust with this big red star, who are we puny humans to argue with it?' Astrologers, all reading from the same sky, would advise weaker kings to be on their guard at such times and look to their defenses. Thus, even through the Middle Ages and into the Renaissance, a cycle of human activity appeared, based on the opposition of Mars, which made kingdoms mobilize and adrenaline, and testosterone, levels rise periodically. This led to the careful study of Mars's movement in the sky, for whoever could predict it best, would get the king's attention and funding. It was in the midst of this pulse beat that the scientific revolution occurred.

Mars, the chief astrological influence between Ares and Scorpio

Mars, at left, assisting the house of Tudor's King Henry VIII and Queen Elizabeth I in center

It can be said that the scientific revolution began with the publication in 1570 of the book *On the Orbital Revolutions of the Heavens* of Copernicus, who was a Pole. Poland stood at the outer frontier of Catholic Europe, a flat land with no natural barriers to movement sitting on the edge of the vast plains of Asia. The broad plains, or steppes, of Asia stretch from the borders of Poland to the great Wall of China. Poland was a nation that had taken a hard course of study in the school of Mars.

In 1241 Poland had experienced an invasion from outer space. The Mongols, an unknown tribe from near the border of China, suddenly appeared and destroyed Kiev in the Ukraine, and slaughtered all its inhabitants. The Mongols then charged into Poland. Using an army of horse-mounted archers and employing gunpowder to create confusion, the Mongols crushed every army the Poles could send against them and destroyed several major cities in the land. Finally they destroyed an allied army of Germans and Poles at Legnica (also known as the Battle of Liegnitz) before charging down into Hungary and ravaging that land as well. Then, as quickly as they had appeared, the Mongol armies disappeared back into the steppes. All of Europe reacted in terror and fear at he reports of what had happened in Poland and Hungary, and armed themselves and waited. It was a good time to be a student of Mars. Fortunately for Europe the expected major Mongol invasion did not come. The Mongols raided several times but the Poles and Hungarians, now being better equipped and forewarned, put up furious and costly resistance. Many speculate, however, that the armies of Europe would not have been able to stop a determined and forceful Mongol invasion. Europe was indeed fortunate. The Mongols conquered China and destroyed Baghdad in the same time period, creating pyramids of human heads everywhere they went. It is believed that their reign of terror was one of the worst episodes of mass murder in history, only exceeded by those in modern times.

Polish Cavalry Monument

The Mongols, everybody's enemy

The Mongols besiege Baghdad before utterly destroying the city

Therefore, Poland was a flat land stuck in the rest of the Middle Ages between the warlike and aggressive Germans on the west and the raiding Mongols on the east, who continued to raid. The Poles, playing off one side against the other for centuries finally emerged as a powerful land empire, with a cavalry force of legendary skill and courage. The Polish cavalry learned to fight both the Mongols and Germans effectively and thus Poland's peace was preserved. It can be said that the Poles, at length, achieved the Dean's list in the school of Mars. This allowed Poland to flourish and created great centers of learning.

Copernicus was a good Catholic and belonged to an institution then immersing itself in ancient Greek and Roman learning. Being a Pole, Copernicus stood at the frontier of European thought, and could gaze into the vast darkness of the steppes. Perhaps this sense of confronting a constant unknown made him bolder than other Europeans of the times. He revived a concept of the ancient Greek philosopher Aristarchus, who said that Earth, together with the other planets, moved in circles around the Sun, and showed that this made the prediction of the movements of the planets simpler mathematically. His system was, in a word, "elegant" compared to the then accepted system of putting Earth at the center of the universe with the planets, Sun, and Earth's moon moving in complex "epicycles," circles within circles, around it.

This heliocentric view looked nice on paper. It held to the "minimum of entities" rule. However, it had the unfortunate problem of demoting humanity from being the ruler of the center of the universe, to being rulers of just one of the planets. This meant that the Catholic Church, Copernicus's employer, was also demoted, because it took care of humanity's spiritual well being. The Church, at that time, prided itself on being able to answer all the "Big Questions." Copernicus realized that his new model of the universe might raise a host of new, and difficult to answer, big questions, and that the Catholic Church might not be pleased. Copernicus handled this problem with characteristic Polish genius; he made sure he was dead before the book was published.

A nervous looking Copernicus

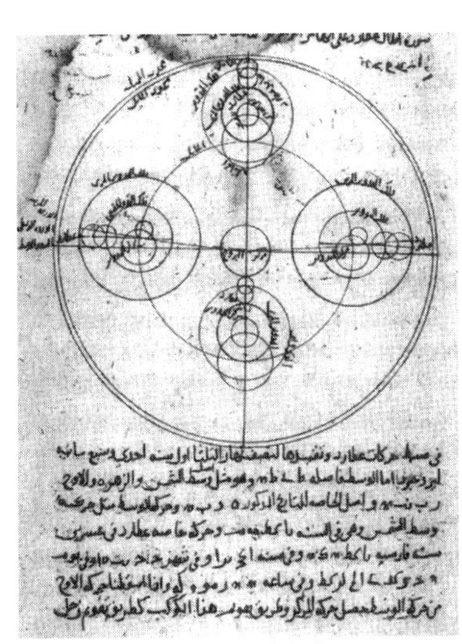

The Geocentric System with its many epicycles

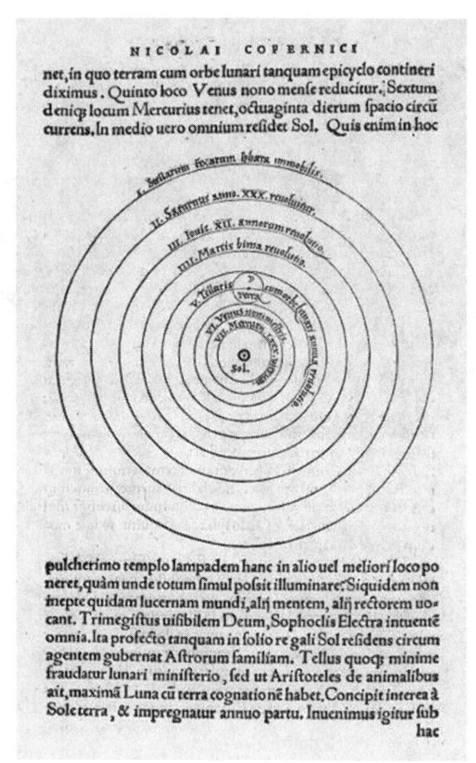

NICOLAI COPERNICI

net, in quo terram cum orbe lunari tanquam epicyclo contineri diximus. Quinto loco Venus nono mense reducitur. Sextum denique locum Mercurius tenet, octuaginta dierum spacio circũ currens. In medio uero omnium residet Sol. Quis enim in hoc

pulcherimo templo lampadem hanc in alio uel meliori loco poneret, quàm unde totum simul possit illuminare? Siquidem non inepte quidam lucernam mundi, alij mentem, alij rectorem uocant. Trimegistus uisibilem Deum, Sophoclis Electra intuentẽ omnia. Ita profecto tanquam in solio regali Sol residens circum agentem gubernat Astrorum familiam. Tellus quoque minime fraudatur lunari ministerio, sed ut Aristoteles de animalibus ait, maximã Luna cũ terra cognationẽ habet. Concipit interea à Sole terra, & impregnatur annuo partu. Inuenimus igitur sub

hac

The Copernican System, much more elegant

The Poles stopped the Mongols from invading Europe further, but they were helped in this by events far to the east that had been recounted in Europe by Marco Polo. Polo had reported that the Mongols, at the height of their power after conquering China, had tried to take the farthest land of the Far East, Japan, and failed disastrously, not once but twice. This pair of defeats badly depleted the Mongol armies of men and ambition. These disasters had been due partly to fierce storms that wrecked the invasion fleets but were also due to the valor and skill of the Samurai, who held the Mongols on the beaches until the miraculous storms arrived.

The Mongols attempt to invade Japan.

Samurai

The Samurai were avid pupils in the school of Mars, and worshiped the god of the warriors Hachiman, who was gentrified by Buddhism as the protector of Japan.

Hachiman disguised as a Buddhist monk

Back in Europe, Giordano Bruno, a Gnostic philosopher of the time, became quite excited about the Copernican concept and carried it further. Bruno went around Europe saying that not only did the Earth and other planets orbit the Sun, but that the stars were in fact other suns viewed far away, with planets orbiting them, and with people like us living on those planets. This meant that not only was humanity merely ruler of a planet, but this planet might not be a particularly important planet in the cosmos. He was promptly arrested by the Church and burned at the stake in 1600 when he would not recant this idea. The Church was uncomfortable with the idea of humanity ruling just a speck of stardust, and being one of myriad other peoples among the stars. Such ideas would prompt questions from the people and the Church did not like questions, especially when they could not answer them.

Galileo, in Italy, was a man of vast genius who created a fancy new telescope. He was professor at the university and wanted to gain favorable attention of the head of his department. He had been analyzing how things moved, and developed the idea of Galilean motion and gravity acceleration. He was a showman who reportedly rolled two cannon balls of different weights off the leaning tower of Pisa to prove that they both fell at the same rate and would hit the ground at the same time. He used iron cannon balls because they were a new high tech invention and flew better through the air than the stone cannon balls used previously. He then used this new science and math for describing how cannon balls moved to develop excellent elevation versus range tables for military artillery. These tables were so good they were standard on the Earth until World War One.

Improved artillery tables were regarded with favor by the Catholic Church. The Church was studying in the school of Mars vigorously in those days. It was concerned with protecting Christian Europe from conquest by the Muslim Turks and conquering the land now held by the Protestants in Germany. However, Galileo soon attracted attention of a less favorable kind from the Church. He pointed his telescope at

Jupiter and saw a set of moons like pearls on a string. The church had basically ignored Copernicus's ideas, hoping they would go away. The church teaching remained that the Sun, Moon, and visible planets represented the seven deadly sins and seven heavenly virtues. The gaggle of moons that Galileo saw orbiting Jupiter in a miniature Copernican system suggested, at the very least, that many new sins might await discovery. This would prompt questions. Galileo was therefore promptly arrested by the church in 1610 after his report of strange new objects in the sky. He was charged with "vehement suspicion of heresy." Threatened with torture, he wisely recanted his sightings of strange lights in the sky and was released by the church authorities with the warning: "in the future, Doctor, leave the big questions to us."

Kepler, who described Mars orbit precisely

Galileo, who saw too much

In a contemporary set of events, Tyco Brahe, court astrologer to the King of Denmark, told his assistant Johannes Kepler to "study Mars" if he wanted to understand the orbits of the planets. For, noted Brahe, Mars had the most unusual of any planetary orbit, being difficult to predict even with the theory of Copernicus, which used circular orbits for the planets. In an epic accomplishment of the human intellect, Kepler was able to explain the orbits of the planets as ellipses in 1609, using mathematics originally worked out by the ancient Greeks, and now rediscovered as part of the Renaissance. In one great leap he not only described the orbits of the planets, but showed that the learning of the ancients was required to understand them. He thus validated both the progressive and traditional elements of the Renaissance. In 1619, he followed his earlier achievements with the discovery that the period of the orbits of the planets was proportional to the radius of their orbit to the 3/2 power.

The world had become larger then, to the Europeans, with the discovery of the New World by Columbus in 1492. This expedition was backed by the Spanishwho then led the European invasion of the New World. It was the collision of that part of the human race that had long ago moved west from Africa until it had run into the Atlantic Ocean and been held there, with the other end of the human race that had moved from Africa and then east into Asia and from there to North America, until it stood on the other side of the Atlantic from Europe. Two portions of the human race that had had no contact for thousands of years were now at a collision point. Cortez led an expedition to Mexico from the Spanish base in Cuba. He was a man who, upon landing in an exotic land with an alien civilization, attacked an empire of millions with 800 men. He had more than sheer bravado and machismo on his side. For one thing, Cortez knew he had the best infantry in Europe.

Cortez, his men, his Indian allies, and his Aztec princess in Mexico

The Spanish man of war, the guerro, was the product of the school of Mars on the Iberian Peninsula. Almost a millennium of war during the Christian "reconquista" of Spain from the Muslim Moors had transformed Spain into a society dominated by war and the men who waged it. Daring, strength, cunning, religious zeal, and aggressive opportunism were qualities to be cultivated and admired on the battlefield and off. Hernando Cortez possessed all of these quantities in abundance. He had landed in Mexico, taken a look at the coastal tribes, assessed their inferior technology, bested them in battle, and been given a lovely slave girl as a gift. She, as it turned out, was a princess of the Aztecs, the real rulers of the land, who had been sold into slavery to their vassals, the coastal tribes. She became Cortez's lover, learned Spanish quickly, and informed him of the whole political situation in the land, and of the legend of Quetzalcoatl, a god Cortez resembled. He instantly digested all of this and saw how to take the whole place. For her it was payback time, for him, in stroke of wild genius, it was a golden opportunity. So in 1516, a quarter century after discovering the New World, Spain conquered territories and populations larger than themselves, and became the richest and most powerful nation in Europe.

That a band of such men of war, led by one such as Cortez, would form the spear point of collision between the old world and the new was not an accident of history. Fortune favors the brave, and also leads them to journey far. Mars teaches us that anyone who comes here to Earth across the starry gulf may have gone to the same school as Cortez.

El Cid, hero of the Spanish Reconquista

The Conquistadors in the New World

The Spanish used Mexico as their springboard for invading the Pacific basin, and invaded and conquered the Philippines in 1565. Enriched and supplied from their bases in the Americas, the Spanish seemed all powerful. They turned north from the Philippines, eyeing the rich islands of Japan, but were stopped there utterly. The Samurai waited for the Conquistadors on the beach. The Samurai had what the Aztecs of Mexico lacked, steel, horses, and something else. Thanks to some shipwrecked Portuguese, several years before, the Samurai possessed gunpowder and firearms when the Spanish arrived. So the Spanish trail of conquest ended abruptly. This collision with the Spanish, particularly their attempt to conquer Japan through Catholicism, having failed to do so by force of arms, caused the Japan to retreat into seclusion until the 1800s.

It remained for Newton to finish the great work of understanding the heavens. Once again it was the orbit of Mars, with its markedly elliptical shape, rather than a nearly perfect circle like Earth's orbit, that was the challenge. In a circle, centrifugal force and gravity balance and are constant. The algebra of the Arabs could suffice to solve this. However, in an ellipse, the forces vary constantly. What was needed was a form of mathematics that could solve for things that varied constantly. With a quantum leap in human understanding, Newton published his epic tome: Principia Mathematica, in 1687, where he invented both the laws of gravitation and the laws of motion, but also invented calculus to describe the motions that resulted. In one step, he validated the works of Kepler and Galileo. To the intellectuals of the time, the cosmos was transformed by Newton from an impenetrable mystery to a grand and beautiful mechanism.

In this entire journey from Copernicus to Kepler and finally to Newton, the orbit of the planet Mars was the chief problem to be understood. Thus it can be said that the planet Mars was not only the inspirer of cutting-edge technology, literally, from ancient times, but acted as midwife to the scientific revolution.

Newton, whose gravitation theory explained Mars orbit

Chapter 2. The Dream of Mars

"Someone looked across the starry gulf with envious eyes."

War of the Worlds, H.G. Wells

In 1851, to celebrate the prosperity and advances of Victorian England, a great international exhibition was held. The exhibition, the first world's fair, was the brainchild of Prince Albert, the consort of Queen Victoria. The centerpiece of this exhibition was the Crystal Palace, a remarkable building made of glass panes supported by cast iron girders. The fact that it was a giant greenhouse was because it was the brainchild of William Paxton, a designer of greenhouses, who took advantage of new technologies for mass producing iron and glass, plus the repeal of the glass tax. Not only was the building the first prefabricated structure ever built, but it also established the glass-metal frame design of the greenhouse until the late 20th century. The Crystal Palace symbolized a new age of humanity, and its declaration of independence from old structures of architecture and thought.

Then in 1859 Darwin published his famous The Origin of Species. Applied to Mars, this suggested that life and its evolution was part of the natural process of things and would occur on Mars. This again reinforced the idea that Mars, with its terrestrial-looking environment, must be a living planet.

The Crystal Palace

The Crystal Palace inside

By the 1800s both the seeing power of telescopes and scientific understanding had greatly increased from the days of Newton and Kepler. However, the human race was still primitive in understanding the cosmos. Understandably, it applied a primitive logic to what it saw through its new telescopes.

Mars and Earth to scale, note polar caps on Mars (JPL/NASA)

Humans beheld the Moon, saw from its barren craters and lack of clouds that it was airless and dead, and from this learned an important lesson about the cosmos, that two planetary bodies could be right beside each other and one could support life and the other could not. They then knew of two planetary types: lunar airless ones like the Moon and terrestrial ones like the Earth where they lived. Armed with this new means of classifying heavenly bodies, humanity boldly looked out of its telescopes and began classifying things that it saw. Mars attracted its immediate interest because every 25.5 months it would come close and one could look at it clearly. Venus, the other close planet, was in the Sun's glare when it got close and was in general featureless and cloud wrapped. Mars, however, had a visible surface with interesting areas of light and dark, and had an atmosphere with periodic dust storms. Mars also had polar caps that grew and shrank with the seasons. Applying the new binary classification system they had learned, Mars was immediately put in the terrestrial category of planetary bodies. Mars had a stormy atmosphere and polar caps like Earth, it turned on its axis every 23.4 hours, almost like Earth; it tilted on its axis at 22 degrees, almost like Earth. Presumably, it also held life like Earth.

This feeling was further amplified when we noted that parts of Mars appeared to darken in a wave-like movement as the southern polar cap shrank. This "wave of darkening" whose observation was made possible by the ever improving telescopes, made each new opposition of Mars a time of great scientific excitement. The sciences flourished then, for it was a time when science and other higher culture could be pursued and supported; it was the time of "Pax Britannica."

Under the 63 year reign of Queen Victoria, from 1840 through 1905, Britain experienced the height of its economic and intellectual accomplishments. The British commanded a massive navy, safeguarding trade with a world girdling empire. Presiding over all of this was Victoria. She was a monarch of limited legal power, but enormous influence. She brought to Britain a sense of stability and high manners, and great esteem for the intellect. The status of women was greatly elevated in society, as was to be expected with a woman in charge. Victoria was aided greatly in her sense of authority by the memory of Queen Elizabeth I, a paragon of skill and wisdom in all manner of statecraft. It was as if Athena, the goddess of wisdom, had been enthroned.

Queen Elizabeth I

Queen Victoria at the beginning of her reign

In 1877, during a spectacular opposition, Alsaph Hall, of the U.S. Naval Observatory, discovered two moons orbiting Mars. They had been uncannily predicted a century before by Jonathan Swift, in the science fiction tale *Gulliver's Travels*. Hall named them Phobos, "Fear," and Deimos, "Terror," the two mythological children of Mars the War God. These had been the consequence of Mars affair with Aphrodite. In that same opposition, the Italian astronomer Schiaparelli made the observation of what he called "canali," Italian for channels or grooves. This was immediately translated by the English speaking press as "canals". Other astronomers saw them also.

Alsaph Hall, discoverer of the Moons of Mars

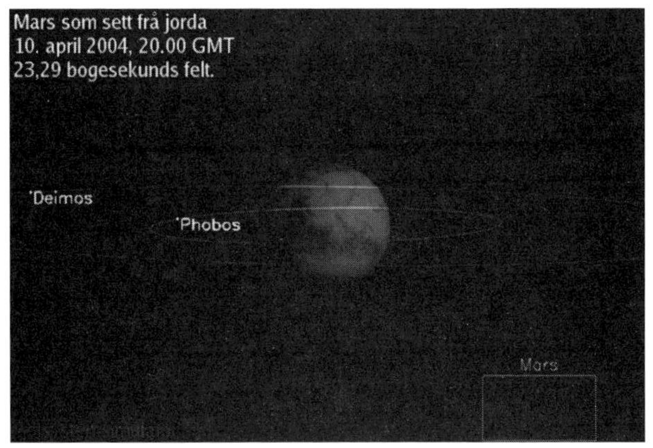

The orbits of Phobos and Deimos discovered by the Naval Observatory (NASA)

Because of the superiority of the human eye over any artificial picture-taking device of the time, and the fact that all observation of Mars had to be made through some portion of the turbulent atmosphere of Earth, the "canali" could not be photographed; they could only be seen and sketched. One would painstakingly look at Mars, wait for the atmosphere to quiet down and the planet to snap into sharp focus, then sketch what could be seen, then stop when the turbulence of the atmosphere resumed its blurring of the planet's image.

Percival Lowell, a son of wealth from Massachusetts, graduated from Harvard with a degree in Astronomy. Breaking off an engagement to a daughter of wealth in 1880, he fled to Japan, then modernizing. Japan had ended its long period of seclusion, induced by its collision with the Spainish, and was actively embracing Western culture and science.

Percival Lowell was entranced by Japan. The farthest nation of the Far East portrayed to him a new world rich and sophisticated yet alien and exotic. He wrote several books on the nation that introduced its culture to America. During this time Lowell became mesmerized by Mars. Perhaps his exposure to a foreign culture so rich and independent to his own got Lowell thinking about exploring an alien world.

Alternatively, as Abe Megahed, a colleague of the author's, has suggested, it was the simple fact that everywhere Lowell went in Japan he was confronted by a flag depicting a large red sphere, which had become the flag of Japan in 1870. For whatever reasons, this rich son of Boston decided to embark on a journey to the most exotic new world of all. In 1894, spurred on by the coming opposition of Mars, that promised to be close and bright, he resolved to build a telescope and see Mars better than any man before him.

Japan somehow inspired Percival Lowell to explore Mars

Inspired by the reports of "canali" on Mars, Lowell invested part of his family's fortune in a superb new observatory. Perhaps attuned to the idea of an alien culture by his years in Japan, Lowell accepted the common interpretation of the canali as a sign of civilization on the red planet. The telescope he obtained was housed in an observatory built near Flagstaff Arizona, where the high, clear desert air made Lowell's the most modern and best located observatory in the world. The air on cold clear nights was remarkably free of turbulence. Lowell took advantage of this to map Mars and name many of its features and regions. He was a poet at heart and gave Mars's regions names from terrestrial antiquity, such as Cydonia, Arabia, and Utopia. The more he looked at Mars the more real the canals appeared.

The flag of Japan depicting the red rising Sun

Noting that Mars seemed devoid of oceans or seas, Lowell conceived the idea of Mars as a planet that was dying and becoming a desert. The canals, he reasoned, were an attempt by a civilization to bring water from the polar caps to the midlatitudes to sustain agriculture. He mistook the sheen of its atmosphere for high density gas, and calculated that a terrestrial range of pressure existed on the planet's surface. Svante Arrhenius in 1903 had discussed the idea of greenhouse gases, which would trap heat in a planetary atmosphere. So a dense atmosphere made for a temperate climate. Arrhenius also revived the older idea of Panspermia, the idea that Earth and other planets had not invented life, but been colonized by microbial spores from outer space. Thus if Earth supported life, Mars must also.

Percival Lowell

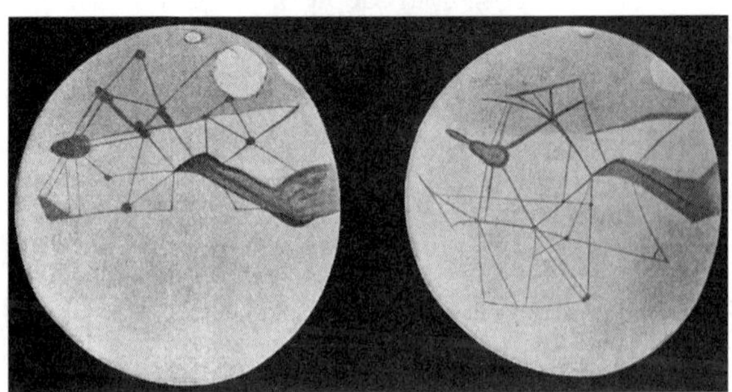

Lowell's drawings of canals

Lowell at his telescope

Armed with observations from what was, at the time, the best sited and most advanced telescope in the world, and encouraged by the ideas current in biology, Lowell wrote a widely read book, *Mars* (1895). This book in turn inspired H.G. Wells to write the horrifying classic *War of the Worlds* (1898), a dark and brutal tale of Victorian England being conquered and destroyed in the same way that the English had subjugated the island of Tasmania and exterminated its natives in the decades before.

Wells drew part of his inspiration from the fantastic tale of the conquest of Mexico in 1520 by Hernando Cortez.

War of the Worlds was notable not just for its portrayal of an invasion from outer space, the use of directed energy weapons and poison gas in warfare, and the idea of biohazards, but for the idea of interplanetary travel, a feature of which was propulsion accompanied by clouds of flaming hydrogen.

H. G. Wells

Konstantin Tsiolkovsky

The mention of hydrogen as the propelling agent of the Martian invasion fleet was prescient of Wells. Konstantin Tsiolkovsky of Russia soon published a widely read book, *The Exploration of Cosmic Space by Means of Reaction Devices* (1903), where the basic mathematics of rocketry were worked out and hydrogen was identified as the most effective fuel.

Mars invades Earth and lays waste to it

At the age of 16, in 1898, the book *War of the Worlds* was read by a young man named Robert Goddard, who found this to be a life-changing event. It led to an epochal vision of the future when, on October 19 1899, he imagined he could build a spaceship that would go to Mars. Studying the works of Tsiolkovsky, he proposed the construction of rockets propelled by liquid fuels and oxidizer, such as nitrous oxide or liquid oxygen.

Robert Goddard

Also springing from Lowell's work was the idea of Mars as a dying world losing both air and water and supporting a doomed civilization. This appears most graphically in the novel *Dejah Thoris, A Princess of Mars*, published in 1912 by Edgar Rice Burroughs. Thus began the long saga of John Carter of Mars. In that novel and the others that followed, Mars is depicted as a planet on life support, its atmosphere maintained by enormous atmospheric plants, who utilize a form of light unknown to humans to create air. Water is carried from the polar caps, a newly explored region on Earth, to water a thirsty Mars in a massive engineering project dwarfing any undertaken on Earth. These novels, along with the War of the Worlds and Lowell's work, created a consensus in human society that Mars was both a living planet and the abode of intelligence.

The first liquid fueled rocket

The sensation created by the *War of the Worlds* demonstrated, however, that the idea of civilization on Mars was regarded with ambivalence. War between different cultures on Earth was a continual fact of life in the late 1800s; would a culture from another planet, especially a planet with desperate inhabitants, behave any differently? Mars, however, soon cast its spell again on the human race in a more ancient form.

In the August of 1914, the light of Mars at opposition shone over Europe. It had been a long hot summer in the Balkans, climaxing when an ardent Bosnian nationalist had assassinated the Archduke Ferdinand and his wife Sophie. The Austrians demanded concessions from Serbia, whom it held responsible, and the Russians, the protectors of their Slavic cousins, had then warned Austria. The Germans had then backed their cousins the Austrians.

As Mars rose brilliant in the east at sunset and ruled the night, Russia mobilized. No United Nations existed to mediate, and to a Europe used to a brief war every generation, a sense of fate and dull-witted endurance settled in. The sense of a head-long plunge into war under the light of Mars was captured by Gustav Holst in his epic symphonic suite, The Planets, specifically, the movement, *Mars, the Bringer of War*. But surely it would be a short war, and perhaps it would settle some issues that had festered for a generation in Europe.

The Great War lasted four heartbreaking years and slew the flower of European manhood. It settled nothing. Every scientific advance the late 1800s had produced was carried to the Temple of Mars and laid as an offering. These offerings were made into weapons to slaughter men on an industrial scale. The horror of poison gas, whose first use against human beings was depicted in Wells's *War of the Worlds*, was unleashed in real life against the troops in the trenches. It was a war of artillery. The Germans, frustrated in their drive to Paris, a drive they had accomplished so easily in the Franco-Prussian war of 1870, built a cannon of unheard of size, termed "Big Bertha," and from their front lines lobbed hopelessly inaccurate shells sixty miles to rain mindless terror on Paris. The conquest of the air achieved by the Wright brothers was also applied to the slaughter. The United States, a cultural child of Britain, formerly preoccupied with developing its massive territories, now entered the war against Germany. However, equally fateful, Russia collapsed and fell under the spell of another red star.

Ares, the Greek god of war, was a proponent of victory by any means. So the Germans, ruled by the Prussian military caste, had employed a form of "thought virus" to take Russia out of the war. Knowing that Russia was feeling terrible social and economic strains, they conducted, in a sealed railcar, the vessel of the thought virus to Russia through their territory and unleashed it. The carefully guarded vessel was Vladimir Lenin. The thought virus of communist revolution spread like wildfire in the suffering masses of Russia, and soon a new political emblem, the Red Star, had appeared in the firmament of the world's thought.

Goddard himself, under the spell of Mars, developed the prototype of the U.S. Army Bazooka, a solid fuel rocket-propelled weapon for infantry, and tested it in the last days of World War I. In the end, despair outran all other factors and Germany collapsed psychologically, and sued for peace. Holst celebrated the end of the terrible war in 1918 with the touching symphony *Venus, the Bringer of Peace*.

The war had exhausted the Western world, and filled it with cripples, widows, and the fatherless. The peace of Versailles that followed it was like the peace settlements that had followed every previous European war. To the discerning eye, the peace set the stage for another cycle of European war in a generation. To the people of the United States, who had waded into the bloodbath just long enough to experience its horrors to the full, the whole episode seemed an utter waste of time and human life. The Americans turned with relief and chagrin to other things. Goddard was thinking about rockets to Mars.

In 1919 the Smithsonian Institution published Goddard's groundbreaking work, *A Method of Reaching Extreme Altitudes*. Goddard proposed using rockets to supplant balloons as a method of obtaining data on the atmosphere at high altitudes. This work was widely translated and read by future rocket pioneers Herman Oberth and Werner Von Braun. However, to an America whose nerves had barely recovered from the Great War and the stupidities that followed it, the idea of shooting vehicles beyond the atmosphere, and even to the Moon, looked like a sure prescription for trouble. A deeply conservative and hermitlike mood had descended on the United States. Shooting a rocket to the Moon might mean, at the very least, having to consult with the League of Nations, the creation of an American president, that America then refused to join. Goddard was attacked by an anonymous editorial in the New York Times for his ideas. In particular his idea of a rocket to the Moon was attacked as impossible since in the vacuum of space the rocket exhaust had "nothing to push against." This attack left him stunned and adverse to publicity. He had pushed the envelope of human thought, and someone nameless and powerful had shoved it back at him. In many countries, reading an anonymous editorial in a prominent newspaper attacking oneself could easily be interpreted as a threat to one's life. Goddard took this as a clear warning. It demonstrated that the spirit of the Inquisition of the Middle Ages was still alive and still determined to preserve the world as we knew it.

Unspoken by both Goddard and the nameless editorial authority, was the idea of a rocket to Mars. If Mars had canals, ran the unspoken logic, sending a rocket there might not be a good idea. It might create trouble.

Goddard was a doer, even if he did not like being an advocate. He quietly applied himself to making his dreams a reality. In 1925 Goddard put his theory into practice and launched the world's first liquid fuel rocket. It used liquid oxygen for oxidizer and gasoline for fuel. It flew for a few seconds and achieved an altitude of forty feet; however, it was the beginning of things that would one day fly to Mars and beyond.

Goddard's work advanced, and he moved his research in 1929 to a place more suitable for his larger rockets, a quiet obscure place: Roswell, New Mexico. He was virtually ignored in the United States, a fact that suited him, but his work was of great interest overseas. The infamous treaty of Versailles had forbidden the Germans from building long range artillery. This was an understandable prohibition after the War's "big Bertha" episode. The Germans thereafter sought in the rocket a means to carry explosive shells to long range without large guns. The Abwehr, German Military Intelligence, therefore sent agents to infiltrate Goddard's circle of associates, and report on everything he was working on. Accordingly, in the small isolated town of Roswell, unknown and unknowable, a series of events and intrigues unfolded in the 1930s, whose consequences would later shake the planets.

Goddard achieved in his researches unheard of speeds and altitudes with his rockets. Fourteen times his rockets of increasing sophistication leapt into the sky, achieving speeds of 550 miles per hour and altitudes of four miles. Gyroscopes, originally developed to guide torpedoes, now guided his torpedoes of the sky. Primitive electric computers regulated the rocket's mechanisms as it flew. High speed pumps, powered by miniature jet turbines, pulled fuel and liquid oxygen from tanks and shot it into a thrust chamber. They fueled fire hotter than a furnace and that shot flame like a white hot dagger far into the air. A thousand different problems were encountered and conquered in this effort. At the peak of the rockets flight, a parachute would deploy, to allow his precious rocket to be recovered intact and refueled to fly again, with all the new lessons learned and applied.

In October of 1938, with war looming in Europe, and with it fears of American involvement, a prankster hoaxed an invasion from Mars. This was the infamous broadcast, in news-broadcast format, of H. G. Wells's, *War of the Worlds*. This prank was authored by H.G.'s ne'er-do-well nephew, Orson Wells. The reaction of some of the public to the broadcast: utter panic, confirmed to the-powers-that-be that Mars and things Martian could be disruptive of public order.

On June 6 1939, the doors of the temple of Mars were flung open once again in Europe. Enlightened Europe entered this war with none of the comforting illusions of the dawn of the Great War twenty years before. In the United States, fathers swore oaths and mothers clutched their sons unto them, and a fierce, if hopeless, resolve settled on the land. In America this was a bitter resolve to stay out of this new war and the circus of European folly it had sprung from. No trench warfare or poison gas was to be endured again by Americans to no good purpose. Europe, Americans thought angrily, would have to settle this blood-feud by themselves. However, the world was too small by that time, and America's place in it too important, for that bitter resolve to endure.

The prankster, Orson Wells 1937

The entrance of the United States into the war after Pearl Harbor, the pummeling we took at the hands of the Japanese in the Pacific and by the German U-boats in the Atlantic, returned Americans mentally to the night of the War of the Worlds broadcast, when all our defenses had been useless. However, America proved an eager and bright pupil in the rough school of Mars, and soon made its presence known in the war, in overwhelming strength. After the protracted and bitter fall of Corregidor, and then the epic battle of Midway, the Americans went on the offensive in the Pacific. In Europe we forced the German Air Force back to defend its own territory. To Hitler, a weapon that could still rain mindless terror on London, even as bombs fell on Berlin, was required. He needed a new weapon that could not be defended against by Allied fighter planes or antiaircraft guns. Werner Von Braun was the head of the German rocket program and had incorporated many of Goddard's ideas into his lance of Mars, the V-2. By the end of that great and terrible war, rockets had shown their ability to rain death on cities from hundreds of miles away in a way that was impossible to defend against. The fact that it could easily carry another innovation of the war, an atomic bomb, made the value of rocket propulsion in the martial realm beyond question. Goddard's adolescent dream to build a rocket to the planet Mars had led, via the fields of Mars, to the ultimate weapon of the great powers.

V-2 Missile

Eisenhower at the Buchenwald Concentration Camp

The Second World War showed humanity two things: one - that its power for destruction and exploration had become greatly expanded, and two - that the human race had limitless capacity for inhumanity. The Holocaust showed that even an advanced and enlightened nation, Germany, could regress to unspeakable tribal barbarism, and turn every skill of the modern age to organize industrial scale genocide.

At the end of World War II, two nations stood all powerful. One had as a symbol the white star, and the other a star of red. Because of Lowell they confronted a planet Mars that had polar caps and canals of water and that both nations esteemed to be a living planet, perhaps holding intelligent life. They also confronted, because of the Wells family, both H.G. and Orson, the concept that life on other planets might not be friendly, and that the possibility of an invasion from another planet had great capacity to incite terror among the population. Also, they now had, courtesy of Mars the war god, the rocket that provided a means to reach the Red Planet.

Chapter 3. The Red Star

"We will bury you."

Nikita S. Khrushchev, Premier, USSR, 1960
(an easily misunderstood statement)

The first of humanity's probes to fly past Mars and return data was Mariner 4 in 1964. Its mission was both to explore the Red Planet and to wave the long lance of Mars the war god. It was Mars the war god rather than Mars the planet that led to the expansion of rocket technology, which in turn led to close range observation of the Red Planet. The Cold War between the United States and its allies and the Soviet Union was at its height and infused every scientific accomplishment with martial implications.

At Mars the red star of the Soviet Union and the white star of the United States dueled for dominance. The duel was personified in two men, Werner Von Braun on the American side and Serge Korolev on the side of the Soviet Union.

Von Braun, our rocket guy

Korolyev, their rocket guy

Both five-sided stars had their origins in antiquity as the Pythagorean pentagram. The star is formed by adding triangles to each side of a regular pentagon. The fivesided star has the remarkable property that the length of the sides of the triangular points are exactly the golden ratio of their bases. Thus to the Pythagoreans the pentagram, and the pentagon, symbolized the golden ratio. The fact that one side would choose white for its star, symbolizing either Christ or Venus, and the other would choose red representing Mars shows how old dualities in human thinking are preserved in modern times. Red had always been a traditional color of power and festivity in Russia, like the suit of Santa Claus. In Russian the words for red, “krassney,” and beautiful, “krasseevya,” are near synonyms. It was natural that the communists would take this color to represent their new order in Russia. That the red color appeared on stars in Red Army uniforms in the revolution of 1917 was apparently due to the communist love of red and the love of all militaries for stars. In communist lore, however, the origin of the symbol is that of a red star plucked from the sky and freed by a hero.

Our star

Their star

The rocket that sent Mariner 4 past Mars was an Atlas, designed to carry nuclear weapons to any part of the Earth, and the precision with which it had flown demonstrated to the world America's skill in aiming any weapon it unleashed. Mariner 4 was launched in December 1964 and flew past Mars in July 1965, shortly after the Cuban Missile Crisis and the Kennedy assassination. The world at large was nervous. However, in the world of science the data Mariner 4 returned was startling.

A modified Atlas ICBM launches the Mariner 4 to Mars

Mariner 4 was aimed at the dark southern regions of Mars that were thought possibly to be covered with vegetation and even canals. What they found instead was a Moonlike landscape devoid of signs of liquid water and covered with craters. This was a shock to many older scientists, who had been raised on the works of Lowell and whose image of Mars was as a desert-dominated but still terrestrial planet.

Careful measurements in the early sixties showed that the atmosphere of Mars was very thin, equivalent to the Earth at 30 kilometers altitude. These measurements were made possible by advances in both technology and the theory of light scattering. Telescopes could both collect more light and better analyze its character, than in the past. It turned out what looked like an atmosphere of Earthlike density around Mars was actually a dusty haze in a much thinner atmosphere. Mars had barely enough atmosphere to allow liquid water to exist on its surface, and then, only at low elevations. The atmosphere contained mostly carbon dioxide, and no oxygen could be detected. However, despite the severe nature of the Mars environment, many scientists remained convinced that Mars was basically terrestrial; they believed that it contained both water and life in some form. This was an extreme extension of the Lowellian concept of Mars as a dying world turning into desert and losing its atmosphere, an idea explored in Edgar Rice Burroughs's *A Princess of Mars* (1933). Things on Mars had simply gone farther "down hill" than Lowell had supposed.

The appearance of craters and the lack of a detectable magnetic field caused the planetary community to move Mars abruptly from the terrestrial category of planets to the lunar category. Planetary science was still laboring under the "either terrestrial or lunar" method of classifying planets and this was amplified by the arrival of a new generation of hard-eyed scientists who wanted to break with the Lowellian past of Mars and confront the brave new truths of the cosmos.

The truth was that the world in those days was haunted by the specter of Hiroshima. The movies about nuclear war, *Doctor Strangelove* and *Fail Safe*, had appeared in 1964; In the former movie the Earth was destroyed utterly. Already by the mid 60s the nuclear stockpiles of the United States and

Soviet Union were large enough to reduce the planet Earth to a cratered and lifeless orb. The Sedan nuclear test had in 1962 created a crater similar to those seen on Mars. The fact that the world saw Mars cratered and barren and immediately leaped to the conclusion that it was completely dead, can be excused as a projection of our own fears.

Hiroshima after the atomic bomb

Doctor Strangelove (1964) where a man rides a nuclear weapon to its target and dooms the Earth

In a curious reversal of this new idea of Mars as dead, a researcher named Sinton, using improved telescope technology, discovered that the dark areas of Mars displayed infrared absorption bands corresponding to organic matter. He had found three such bands, but two had been explained by the interference of heavy water in Earth's own atmosphere. The third band in the infrared at 3.4 microns, associated with methane and other organic molecules, remained unexplained, but such was the character of the times that its meaning as a life indicator was dismissed. The incident appeared to reveal a new feeling in the scientific community and the government that funded it; they were hostile to life on Mars. This hostility appears to have been due to either the panic associated with the War of the Worlds radio broadcast in 1938 or the public unease over the UFO phenomenon starting in the late 1940s.

Mariner 4 (JPL/NASA)

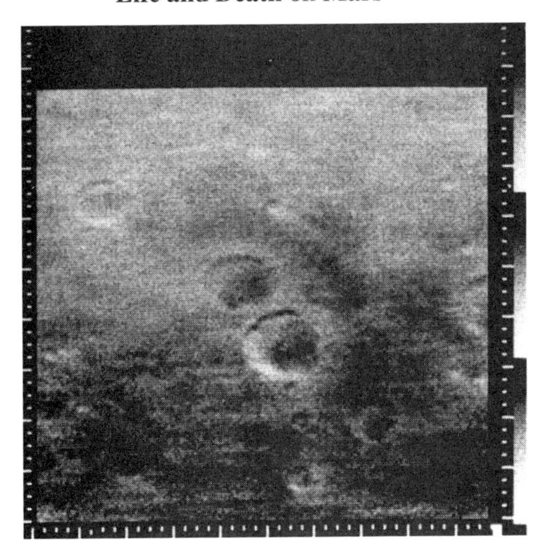

Mariner 4: no canals (JPL/NASA)

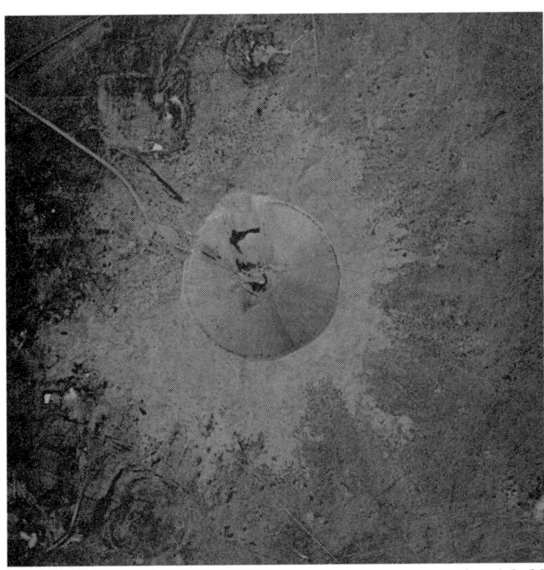

Crater formed by the Sedan nuclear test in 1962

It can be argued, that if the government knew that some UFO phenomena represented extraterrestrial intelligence and decided to conceal this from the public, then such a policy would have a chilling effect on all searches for extraterrestrial life or intelligence. This follows from the fact that the federal government funds most scientific research and thus sets its priorities. The policy need not even be overt but instead implicit, making the search for life respectable, but a positive outcome forbidden. The search for extraterrestrial life would then become a government-sponsored Victorian romance, long on sighs, flirtations, and fond looks, but never destined to be consummated. Whatever its origin, this apparent official hostility to exobiology led to increased ambivalence in the 1960s to the obligatory search for life on Mars in 2010. For the scientific community the search for life on Mars had become a search for life they did not want to find.

As a new planetary alignment arrived each 25.5 months more probes were sent to Mars by both the United States and the Soviet Union. Space had become a great stage on which national technological prowess was displayed, akin to the Olympian games of ancient Greece. In this sense it was a stabilizing and distracting exercise in the Cold War, while back on Earth small wars broke out in the Middle East and Indochina. Both sides demonstrated enough technical skill in space that neither wanted a real armed confrontation.

Mars slowly began to yield new mysteries as the amount of probe data increased. The shock of finding craters had now worn off; what was apparent was that the number of craters on Mars surface was much lower per square kilometer than that found on the Moon. Mars sits on the edge of the asteroid belt in the Solar system, the source of many meteorites on Earth. Mars, being closer to the asteroid belt should be hit with more meteorites and thus have more craters than the Moon per square kilometer. That is, Mars should have more craters than the Moon if its history was lunar, with no air or water erosion since 4 billion years ago. It was the suggestion of Carl Sagan that some powerful force of erosion had been at work to wipe out the craters that had formed over time. Earth has few visible meteorite craters; they have been erased by an array of erosive forces. The forces include volcanism and wind, but the most important force is liquid water. Thus the idea was proposed that while Mars was a cold, dry, nearly airless desert now, perhaps its atmosphere was much warmer and denser in the past and this allowed water to exist and move on its surface as it did on Earth. Thus the idea of Mars with a terrestrial past or the "Ancient Eden" idea was born.

The planetary opposition of 1969 found a fleet of human space craft in orbit around Mars. The Americans had launched Mariner 9 and 10 to Mars, but 10 had ended up in the Atlantic off Florida. Mariner 9 however, had arrived at Mars and achieved orbit successfully. Not to be outdone, the Soviets had launched two probes to Mars each with an orbiting space craft and a lander package. The Russians were attempting to achieve a technological triumph at Mars, which after all was the red star, their chosen symbol.

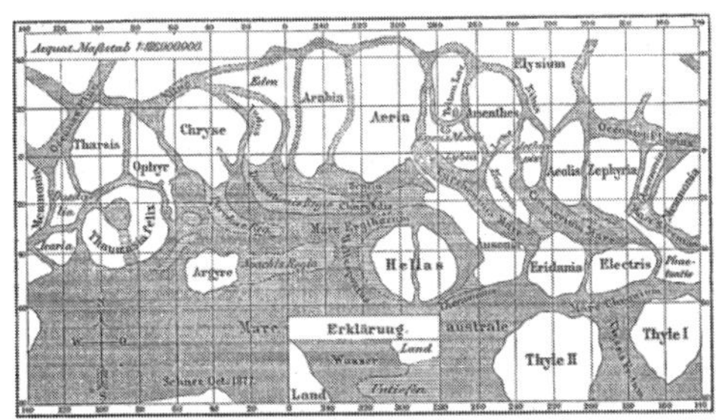

What some scientists expected at Mars before Mariner 4

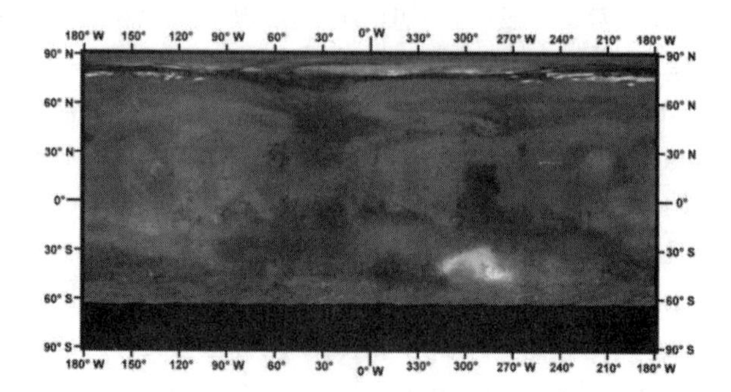

What was found on Mars (JPL/NASA)

Fate moved its hand, however, when Mars was struck by a global dust storm

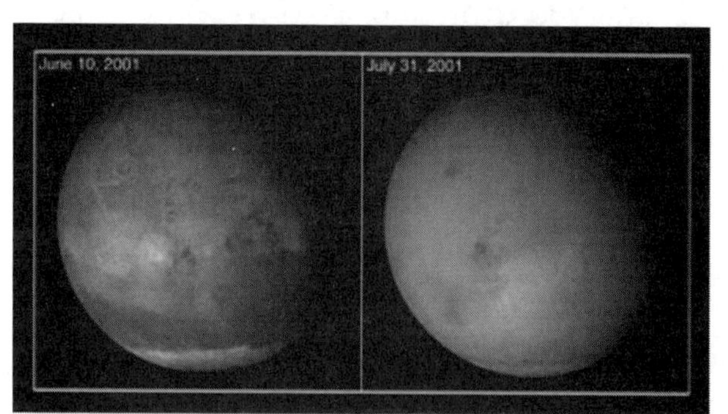

Mars before and during a global dust storm (JPL/NASA)

The Russian Mars probes failed in the face of the great dust storm, because their mission was preprogrammed and could not be changed from Earth. The two probes simply snapped images of the dust clouds then released their landers into the teeth of the storm. Only one lander sent back a few seconds of transmission before it was apparently dashed against the rocks as it was dragged by its parachute.

The Mariner 9, however, following commands from Earth, rode out the storms in orbit in a dormant state and awoke when the storm subsided. Through the clearing atmosphere, Mariner 9 discovered a Mars completely different from what most scientists had expected. Orbiting Mars in a polar orbit, Mariner 9 imaged the entire planet, not just the south, and discovered that Mars had a split personality, or dichotomy, in terms of landscapes.

Mariner 9 (JPL/NASA)

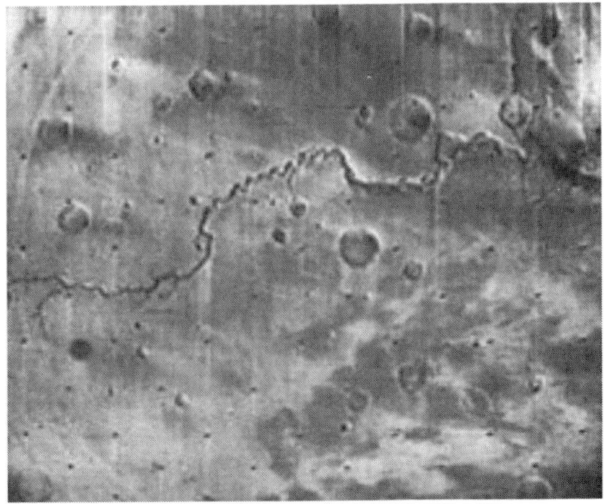

Old river bed from Mariner 9 (JPL/NASA)

The north-south dichotomy of Mars is now understood to be one of the planet's most important features. The south of Mars is heavily cratered highlands, many kilometers above what is considered the mean elevation of Mars. The pattern and density of craters looks very similar to the Highlands of the Moon, and is believed to be nearly as ancient. The north of Mars however, is dominated by smooth plains at much lower average elevation.

Separating the two hemispheres at the equator is a vast fissure or canyon that opens at one end onto the low plains and at the other end breaks into a web of crevices called the labyrinth. At the end of the labyrinth sits a massive plateau crowned by the tallest and largest volcanoes seen in the solar system. Mars was obviously a planet where massive forces of geology, earthquakes, and lava had a festival throughout the past.

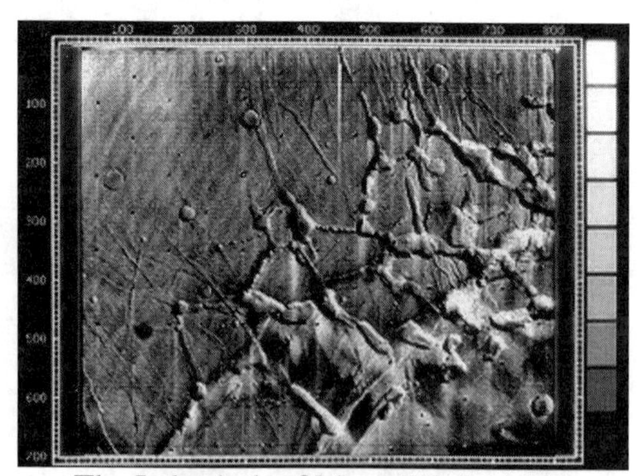

The Labyrinth of Mars (JPL/NASA)

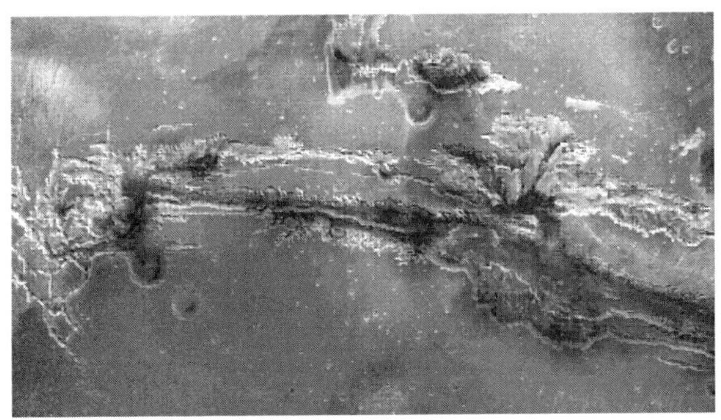

Vallis Marineris (JPL/NASA)

On the boundary of the northern plains system Mariner 9 found the entrance to a large system of parallel canyons, running predominately east-west for 3000 kilometers. This great tear in the surface of Mars ends in a spiderweb of cracks in the Martian surface called the Noctis Labyrinthus, the "Labyrinth of the Night." Beyond the labyrinth is the Tharsis uplift, a vast Plateau on which four enormous volcanoes sit. The volcanoes are arranged in a row of three across the path of the canyon system like chess pieces, with the largest of all the volcanoes, Olympus Mons, standing behind like the queen on a chess board, guarded by her pawns. The origin of the canyons is believed to be a frustrated form of plate tectonics related to the stresses that created the Tharsis uplift. It is as if the massive geologic forces unleashed by the canyon had failed in their attempt to create a new continent, and thus relief of this frustrated pressure was the sprouting of the massive volcanoes.

The Vallis Marineris, named in honor of Mariner 9, dwarfs any similar feature on Earth. The Grand Canyon in the United States is 2 kilometers in depth at its deepest, 16 kilometers wide on average, and 400 km long. The Vallis Marineris is 8 km deep, on average, and 75 km in average width. It could thus hold four grand canyons stacked on top of each other and ten lined up end to end.

Like the Grand Canyon on Earth the walls are of russetcolored stone in thousands of layers that run for hundreds of kilometers. In some places five kilometers of tightly spaced layered rock can be seen running from the basement rock to the top of the canyon. In the Grand Canyon 1.6 kilometers covers half a billion years of history, back to the pre-Cambrian. The rocks of the Grand Canyon walls are also russet, because the sedimentary rocks were laid down when Earth had developed an oxygen atmosphere. At the very bottom of the Grand Canyon the rock turns black, because it was laid down before the era of oxygen. This rock layer is called the Vishnu schist.

In the walls of the Grand Canyon on Earth is told the last portion of the story of life on Earth as well as the story of its climate. This can be traced by reading the ages of the rocks using nuclear methods. The Earth is believed to be 4.5 eons old, an eon being a billion years. The whole cosmos is estimated to be 13 eons old, so the Earth was born in the last third of the cosmos existence.

The Grand Canyon of the American Southwest

Once a great blue star we shall call Titania shone near here, so large and hot it could not die a quiet death. The star that burns twice as bright will last half as long, so it ran through its supply of hydrogen in a few million years, converting it to helium and energy. Soon it was burning helium into oxygen and carbon in its inner core to keep its light shining brilliantly. Even so the helium began to run out so it burned the oxygen also into silicon, and before long Titania had spent the glory of her youth and grown tired red and swollen. In the core of the star the end of a long road of nuclear alchemy was reached; the core was now iron, and no more could the core sustain the energy flow that had poured from the star's surface. The star collapsed on its dying core, and crushed its central plasma to such density that the thermal radiation created matter and anti-matter. The outer layers of the star detonated and consumed all their remaining energy in a few seconds before exploding into blinding blue light that outshone the whole galaxy before it faded. So Titania died and her body became a gossamer expanding cloud rushing outward into the surrounding stardust.

The gas and dust of Titania's brilliant death, still simmering with radioactive heat, formed the Sun and its planets and many other systems as well. Her death was deceptive, for actually she was pregnant with the stuff of new stars and planets.

The formation of the Earth, Mars, and Sun, together with the rest of its family of planets was apparently triggered by this supernova explosion. This explosion gave all the planets radioactive elements as well as stable ones, and the decay of these elements into "daughter" elements allows us, by measuring the ratio of parent to daughter element present in each mineral portion of a rock sample to measure its age. This method of reading the nuclear "clock" in each mineral sample from a rock can only read back to the last time the rock melted or crystallized out of lava or water. Such melting or crystallization "re-sets" the clock and can date not only the formation of the rock, but also heavy shocks or heating. Using a variety of parent-daughter nuclear systems involving different elements, an expert rock geologist can unravel even complex stories of the ages from a simple rock that most people would hardly notice.

From other investigations on Earth we know that life began very early, with microfossils and other preserved organic residues found in rocks 4 eons old. Some say that life arose on its own, from some primordial soup. However, it is generally agreed that even a so-called simple-one-celled-animal capable of respiration, ingestion, and excretion, not to mention reproduction, is so complex that the chances of it appearing spontaneously are very improbable. Life appeared quickly, and this has led to the revival of another idea due to the Swedish scientist savant Arrhenius. He proposed that life predated the planets, and had arrived here as spores from space and thrived as soon as Earth cooled enough to have oceans. This idea is called "Panspermia." However, it only explains why primitive microorganisms could have had an early start on Earthwith. Many other things have happened.

The earliest microfossils were just that, primitive one-celled animals. What is interesting is that these simple one-celled animals continue to be the sole form of life found in the rocks for several eons. It was as if life on Earth, having become established, saw no need to advance and was content to remain primitive. However, this comfortable situation was not to endure. One problem was photosynthesis, which some bacteria had either evolved or had arrived with. This process used sunlight to make energy for the cells to live on, but like some greedy industrialists, they were polluting the atmosphere with a toxic waste product of photosynthesis, called oxygen. The gas was so reactive it was killing most of the bacteria that had peacefully coexisted with the photosynthetic microbes before; it was also turning the rich dark basalt rock bright red. Oxygen was not a good-neighbor molecule.

In the midst of this crisis of rising oxygen concentration one half billion years ago or perhaps because of it, a sudden explosion of biological diversity and evolution occurred. Whether because of an increased mutation rate due to increased oxygen, or the innovation of bacteria that could not only tolerate oxygen but use it to burn sugar and fat in a supercharged metabolism, the world was changed forever. New and horrid phenomena appeared: in an ocean full of bacteria happy to eke out a peaceful existence subsisting on mildly energy rich compounds spewing from undersea volcanic vents, a new strain of bacteria appeared that ate other bacteria, and burned their bodies with toxic oxygen in order to rapidly pursue yet more sedentary bacteria. To the swift, the ocean was one big buffet, and the swift burned oxygen. Predation had come to Earth.

Soon an arms and technology race was in full swing in the ocean, every living thing was learning to breath oxygen to either eat somebody else or else wiggle more quickly to escape from being eaten. Bacteria were banding together to prey more effectively; others formed defensive communities where some cells actually produced tough mineral shell material for defense of the colony in return for being fed by the other cells. Other cells grouped together to attack these defensive colonies with more and more effective toxins and pinchers to pierce through the shells, so they could get at the juicy cells hiding inside. A cornucopia of strange and bizarre experiments of nature is recorded in the rocks, animals with names like “hallucinogens.” While this bizarre episode lasted for a brief period of geologic time, the ecological organization of predator-prey, plant photosynthesis-animal respiration has survived until the present; only the scale and complexity of the players has changed.

In the Grand Canyon of Earth, a major cross section of geologic history is displayed in the walls. The account begins with the Precambrian, and continues though the different ages of the Earth, the Silurian, the Carboniferous, the Mesozoic, and Cenozoic. In the layers of the Grand Canyon is recorded the awful ravages of the Chixelube impact that wiped out the dinosaurs and left a crater 180 kilometers across. It appears that the Vallis Marineris similarly displays the ages of Mars, but the events it records can only be guessed at.

Much of Mars history can only be estimated. It is commonly believed that the Vallis formed in the middle of Mars's geologic history, approximately 2 billion years ago. A crude estimate can be obtained by using the depth-to-age ratio of the Grand Canyon, so that the Vallis Marineris's exposed strata represents perhaps 2.5 billion years of climate history.

The very bottom of the Mars canyon has areas of very dark rock like the Grand Canyon of Earth. This rock has been revealed to be olivine, an emerald green rock.

What Mariner 9 saw in the details of its images was even more astonishing than the grand vistas it beheld. Everywhere on the landscape of Mars were signs of past erosion due to some fluid. Dry water channels, enough to make even Lowell and Schapparelli smile in their tombs, were everywhere on Mars, amid the numerous craters of the ancient highlands and threading across the smoother landscapes of the north. That the fluid that carved these channels was water was immediately grasped by most scientists. Water, composed of hydrogen, the most abundant element in the cosmos, and oxygen, the third most abundant, is itself the most abundant chemical compound in the cosmos. No other substance that was both massively abundant and also liquid under reasonable past Martian conditions could be imagined. Thus, scientists had barely formed a consensus about Mars past, that it was lunar, before the avalanche of Mariner 9 data caused the lunar consensus to be thrown into doubt. If there had been flowing water on Mars, could there not have been life as well? But was was argued the channels were too transient.

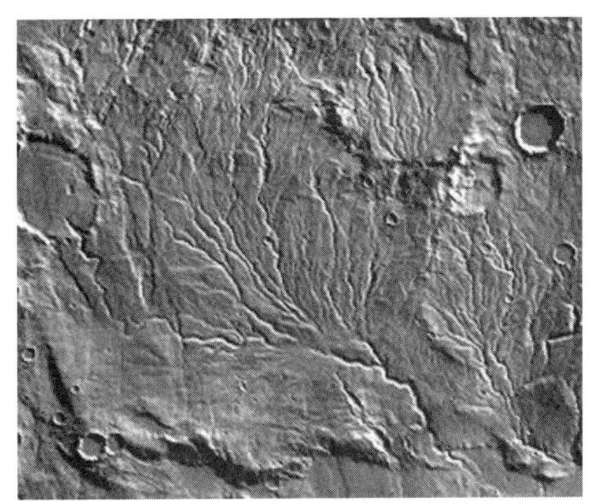

The "canali" of Mars (JPL/NASA)

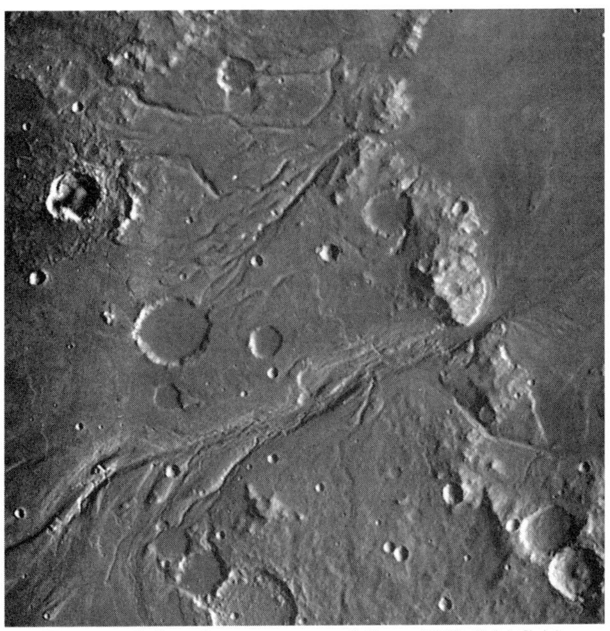

Former lakes in craters on Mars (JPL/NASA)

So Mariner 9 had discovered a new Mars with a watery past. But this sounded too much like the Mars of Lowell, and Mars had already been assigned to the Moon-like category, so the importance of the water channels was minimized.

Life and Death on Mars

Chapter 4. The Vikings of Mars

"The Viking landers have more to do with Lowell than with modern Mars science."

NASA official, June 1976

To address the life question the Viking probes were sent to Mars and arrived in July 1976. The Vikings were meant to display not only the technological prowess of the United States; they were to be part of its bicentennial celebration as well. The Soviets had by this time given up on Mars, and were concentrating on Venus, a closer and easier target where they could control the mission in real time. The Viking probes, A and B, representing the skill of NASA's Jet Propulsion Laboratory at its height, assumed orbit around Mars. The Viking probes were composed of two parts, an orbiter to take images and perform other remote measurements from orbit, and a nuclear-powered lander to set up a scientific station on the surface to measure Mars's atmosphere and its soil properties. The other purpose of the landers was to bring samples of the soil into a miniature biology laboratory. There, using ingeniously designed mechanisms, the soil samples would be tested to see if any living organisms resided inside them.

The Vikings see Mars. (JPL/NASA)

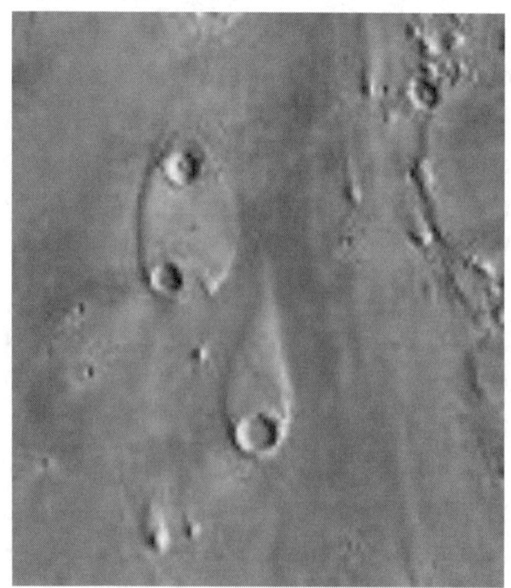

Former islands in a Mars flood (JPL/NASA)

A whiff of danger surrounded this Viking mission. Not only was it costly, several billion dollars, but Russian surface probes had crashed while attempting to land on Mars, so the planet was fully capable of destroying these probes as well, if the mission controllers were not very careful. Thus the spacecraft assumed orbit, and imaged the candidate landing sites at much higher resolution than Mariner 9 to ensure their safety. Only after certifying the safety of the landing sites by images and radar from Earth did the orbiting spacecraft release the nuclear-powered landing craft.

One of the two Viking spacecraft carrying a lander (JPL/NASA)

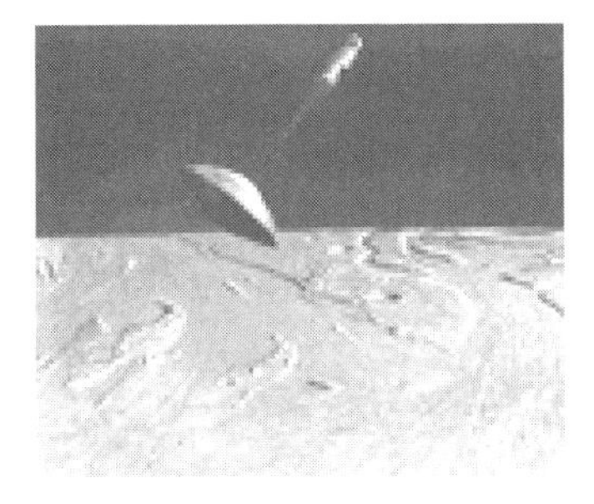

Descent of the Viking Mars lander (JPL/NASA)

However, the danger ran deeper than just a landing mishap. Suppose the lander experiments found life? What then? Would this not provoke a spiritual crisis within humanity, when it realized it was not God's only and most beloved child? We could accept that we were not at the geometric center of the universe, we had dealt with that metaphysical crisis, and moved on. Now we knew the universe had no center, so that geometric demotion did not matter anyway. But we were the center of the biological universe, and the top of the food chain to boot. If they found life on Mars we would truly be demoted, from owning the only oasis in a huge desert to being microbes clinging to a speck of dust. If the Mars life experiments were positive, would this not mean some of the stars were looking back? Were they looking at us, as Wells had suggested, with envious eyes? Would a positive result of the life experiments mean panic in the streets as in 1938?

More urgently, perhaps, would a positive life sign not provoke troublesome questions to the government? Questions like: what are those lights in the sky at night?

The lunar Mars model people, however, knew better than to be apprehensive about the life experiments. They faced the experiments with calm confidence: the life experiments would be negative. They would be negative because Mars was dead as a doornail. In fact, in their opinion, the Viking life experiments were a shameful waste of payload space on the landers. Even testing for life was a chore that had nothing to do with modern Mars science. They knew this with every fiber of their being. They were instead looking forward to the mineral and atmospheric results of the lander probes of the surface. They also waited with anticipation for the Viking images from orbit, in far greater detail than Mariner 9, and in color.

The images of the Viking orbiters verified and expanded upon the profound wonders found by Mariner 9. The channels were everywhere, mingling with craters and emptying into and from them, as if the craters had been lakes full of water. The walls of the Vallis Marineris, were found to be made of sedimentary layers that ran continuously for hundreds of miles, suggesting immediately that the layers had been laid down in a huge lake or sea, as had been similar sedimentary beds on Earth. The tops of the volcanoes were fresh and free of craters, suggesting that the volcanoes had been recently active. The water channels ran across the cratered highlands and then abruptly ended at the edge of the low plains. Some terrains on Mars had so few craters that they looked terrestrial. On some of these nearly terrestrial terrains dried water channels ran sinuously. On others, exposed cliffs revealed millions of parallel layers stretching down for kilometers into vast chasms. To the terrestrial geologist, the message of these images was unmistakable: water in massive amounts had moved and sat in lakes on Mars surface for eons, almost until the present.

The Viking A lands on Mars (JPL/NASA)

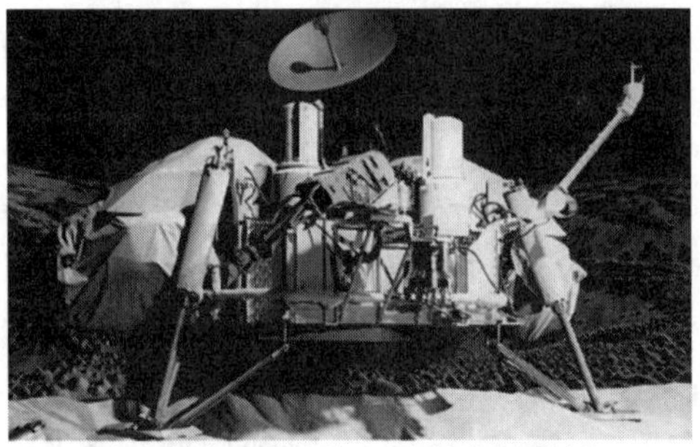

A Viking lander (JPL/NASA)

The summits of the massive volcanoes were imaged in great detail revealing that the volcanoes had erupted recently in geologic times. Mars was obviously a planet that had known a rich full life of every sort of geologic activity known on Earth and this geologic dynamism had continued until recent times.

The determination that some areas of lava flows and water activity were recent had to be made by inspection from afar, at the time the meteorites from Mars were sitting in museums unrecognized. Since they had no known Mars rocks to test for radioactive ages, the Mars scientists were counting craters.

Space is a realm of extremes, of hot and cold and also relative velocities. Under our atmosphere on Earth, things can move fast by human reckoning, as fast as the wind, as fast as sound. However, to move faster than sound, a few hundred meters a second, requires high levels of power. This happens because the air can no longer simply step aside and let you pass by, it piles up in front of you and makes shockwaves and high levels of heat. To go supersonic requires enormous engines; to go hypersonic, many times the speed of sound, requires a large rocket engine. In order to achieve orbital velocity a satellite must be boosted to hypersonic velocities, to stay in orbit at this velocity, it must be above the atmosphere, in the vacuum of space.

Gravity scoops up all the gas in space and piles it up as atmospheres on the surface of planets, but beyond the tops of those atmospheres exists a vacuum of the most extreme purity. In that vacuum, things move effortlessly and without power being required. What objects moving in space do is contain power. Planets, moons, and everything else moves in the vacuum of space like a vast well-designed machine without friction; that is, until something collides with something else. With nothing to slow down objects relative to each other, these occasional collisions occur at hypersonic velocities. On impact the smaller of the objects will turn instantly into blue-white-hot vapor and create an explosion of energy many times that from an equal mass of TNT. For objects of even everyday size coming in from space and not being burned up in Earth's atmosphere, the results dwarf human experience. The Barrenger meteor crater in Arizona, for instance, was caused by the impact of a mass of nickel-iron alloy the size of a semi-truck. The Earth is full of such craters, but only the most geologically recent ones show themselves clearly; the older craters have been eroded away by water, wind, and lava. To see what the Earth would look like without these erosive processes at work, one need only look at the full Moon.

The Moon was where the methods to relate ages of terrains to the number of craters they contained were worked out. The Apollo astronauts had gone to the Moon six times by 1976, and had brought back nearly a ton of rocks from its surface. The rock samples had been gathered at several carefully selected sites of the lunar surface. The rocks had been carefully selected by the astronauts and documented as to where they had been found. The ages of the rocks had been determined by radiometric methods back on Earth, where it was found that the average ages of rocks from each site could be correlated with the number of craters above a certain size per square kilometer of the terrain at the site. These methods found that the rocks from the rugged highlands of the Moon, the most ancient of the moon rocks, were 4.5 eons in age. These rocks were older than any found on Earth, and dated from the foundation of the solar system. The lunar highlands, where they were found, are areas with very heavy cratering. This heavy crater pattern had a very chaotic set of statistics, meaning objects of a wide variety of sizes had crashed at equal rates into the lunar surface. Younger rocks from the Moon, however, dating from 4.0 eons age to 2.0 eons, were found in the dark smooth areas of the Moon, the maria, that had been flooded with lava.

The lunar Maria, such as Oceanus Procellarum, the Ocean of Storms, showed a much different pattern of crater statistics than the Highlands. Instead of reflecting the primordial chaos of the foundation of the solar system, the pattern of size distribution versus number per square kilometer showed an orderly statistical pattern with small craters being more frequent than large ones. The logarithmic progression of these crater statistics on the Moon indicated that, once things in the solar system had gotten settled down, cratering proceeded at a fairly uniform rate of the eons, with the same distribution of sizes of impactors eon after eon. This was very useful, because if the rate was the same over the eons, you could count craters from orbit,, check ages of corresponding rocks on the Moon, and thus estimate the ages of the terrains that the craters occupied.

Ages on the Moon meant not the actual age of the surface, but the age since the last episode of lava flows or the last nearby giant impact had covered the surface with debris and rendered it smooth and free of craters. Therefore, the Apollo missions had given humanity a new tool for measurement, a book listing crater numbers per square kilometer above a certain size where also rocks of so many eons had been gathered. By extrapolating these correlations of rock sample radiometric ages to crater counts, scientists could estimate a rate at which craters above certain size formed on Earth or other nearby planets. This meant you could estimate geologic ages of landscapes on the Moon by taking images and counting craters on the landscape.

This lunar crater-chronology method was immediately applied to Mars. This was done despite the fact that Mars was closer to the asteroid belt, and should have a higher rate of cratering. Not surprisingly, the ages found on Mars were lunar, and for a brief moment, the Mars-is-like-the-Moon party rejoiced. They argued that the signs of liquid water in Mars's past must date from some early period during the intense bombardment, when blocks of ice were melted under the Martian surface and water would flow for a short period before freezing and being lost to space. After this chaotic period, the lunar Mars party advocated, Mars had settled quickly into a comfortable lunar quiescence. However, numerous problems for the lunar-Mars advocates surfaced in the new Viking images.

The new images of Viking were crisp, startling, and covered large amounts of the planet's surface in great detail. In the end it would take 50,000 images covering the whole planet's surface. The images showed objects easily thirty meters across, and sometimes 10 meters across. They showed water channels on terrain with few craters, as well as great floods emerging from chaotic broken terrain that rushed apparently headlong into the smooth plains and then disappeared. They showed networks of small channels feeding into larger ones like a rain-fed watershed network found on Earth. On the flanks of volcanoes, the images captured patterns of drainage channels nearly identical to those found on similar volcanoes in Hawaii. The images showed great fields of sand dunes surrounding the northern polar cap that looked like those on Earth. This led to an important question: Was the rate of cratering on Mars the same as on the Moon? The answer lies in the construction of the Solar System.

In the Solar System, the inner planets close to the Sun are small, rocky, and with distinct surfaces. Mars marks the farthest outpost of this inner solar system from the Sun. Beginning at Jupiter, the next planet outwards, the planets are large, gaseous, and with no solid surface. In the no-man's land between these two realms of the Solar System, between the orbits of Mars and Jupiter, lies the asteroid belt, a disk of circulating rocks ranging from pebbles to Moon-like Ceres with a diameter of 500 kilometers. No large planet formed there, or possibly two formed and shattered each other in a collision. In any case the gravity of Jupiter constantly stirs the asteroid belt, frightening them into rings separated by the Kirkwood gaps, and slowly driving the asteroids of the belt into the inner and outer solar system.

Mars is closer to the asteroid belt than the Earth and Moon; in fact, it can be said that Mars defines the inner edge of the asteroid belt. Asteroids of all sizes are being driven inward toward the Sun by Jupiter. If they get past Mars they hit Earth eventually. The asteroid belt is thus the source of most meteorites that fall on Earth. Their orbits can be calculated back to the asteroid belt by tracing the fiery tracks meteorites make when they enter Earth's atmosphere. However, this whole arrangement makes Mars the target for most asteroids fleeing Jupiter's influence. This means the rate of meteors hitting Mars is probably much higher than those reaching the Earth and Moon. Thus, argued some, using the lunar crater chronology would give wrong ages for Mars. If the cratering rate was higher than the Moon, then the ages of everything on Mars would be younger than estimated by the lunar method.

The Mars community solved this problem in a creative way. Headed by the eminent Mars scientist Charles Hartmann, who is a talented painter and an author as well as a scientist, a group proposed a table of ages for many regions and features on Mars, where the cratering rates were considered to be 1x lunar, 2x lunar, and 4x lunar. The 2x lunar rate was considered most likely, given Mars's proximity to the asteroid belt, and gave a set of ages that showed Mars to be a dynamic planet until two eons ago.

If the ages of the water channels on Mars were younger than supposed, then this would mean that Mars's climate was earthlike for a large portion of its history. The liquid state is the most precarious of all states; gases and solids can exist in a broad range of temperatures and pressures, but liquids cannot. This property of liquids is especially true of water. If liquid water existed somewhere, a narrow range of temperatures and atmospheric pressures must have existed in that place. Accordingly, liquid water defines the Earth environment in terms of pressure and temperature and therefore defines the only known environment that supports life.

On July 20, 1976, in orbit around Mars, the American Viking Lander 1 fired its explosive bolts and was freed of the Viking 1 mother ship. Falling from orbit, it lit up the russet Martian sky like a meteor as it fell and its heat shield turned white hot from the friction of the hypervelocity probe with the atmosphere. A mile above the surface it jettisoned its heat shield and a drogue parachute deployed followed by the larger main parachute. It then slowed in the thin Martian air, which was fortunately not disturbed that day by a dust devil near the landing sight. The lander radar probed the onrushing ground below. Fifty meters from the ground the computer signaled that a firing solution had been found. The parachute lines were severed by an explosive charge and on pillars of flame from its rocket engines the craft descended the remaining distance and then settled to the ground. For the first time, a human creation had landed intact on another planet. It had landed in Chryse Plantia, "the fields of gold" at the place where three former water channels had emptied in times past.

The craft immediately went to work pleasing its ecstatic controllers at the Jet Propulsion Laboratory in Pasadena. It took images of the Martian landscape, showing it looked like desert in the American southwest. Strangely the exact color of the sky at various times of day has provoked debate until this day, for the methods used to determine the right color balance in the images were also subjects of debate. It was an omen of things to come, for the results of the major experiments carried on the probe were to provoke a debate that rages until the present.

Soon after the dust had settled at the landing site, the probe scooped up some soil and carried it inside. The soil was to be tested for signs of life. Three tests were performed: The Gas Exchange Experiment, which looked for respiration by possible microbes in the soil; The Labeled Release Experiment, which looked for ingestion of radioactively labeled nutrients; and The Pyrolytic Release Experiment, which looked for signs of bacterial growth using radioactively labeled nutrients. To make a clear differentiation, half of each sample was heat sterilized to kill any Mars bacteria and then tested also.

Based on the protocols worked out on Earth in comfortable conference rooms, before the Viking probes were launched, the life tests were positive. But the lander was on the cold plains of Mars while, in the russet sky, two moons named Fear and Terror moved overhead. The results were not tidy, not like Earth-life. Weird chemistry was going on in the life experiments. Only the Labeled Release experiment was giving results precisely as predicted for life, and the protocols called for two out of three experiments.

The human race was suddenly poised at a precipice. To accept the verdict of life, based on the experiments on Mars, would create a firestorm on Earth. The human race would be demoted from being the master of the center of the biological universe to being the occupants of a speck of dust, and not a particularly large speck of dust. Many questions would be asked also of the powers that be, and this would lead to trouble. To plunge humanity into such a crisis required unambiguous proof of life, not just evidence.

The samples were put in a gas chromatograph on board the lander and cooked to see if any humus existed in the soil. They could find none. In fact the whole Martian surface was being sunburned by harsh ultraviolet light from the Sun, with no ozone layer or thick atmosphere to mitigate it. Not only would most Earth life die instantly under those rays, but even the organic molecules that made up the dead organism would be hammered to pieces by the photons of that light, sunburned, or photolysized into carbon dioxide and water. Despite the fact that the life experiments were far more sensitive to living things, since things could grow and multiply from a single spore in them, and the fact that the gas chromatograph was never proposed as a test for life when the mission was planned, the gas chromatograph results were invoked to bring closure to the debate. The verdict was that no life had been found on Mars. Two men who were part of these councils objected strenuously to this opinion: Robert Jastrow and Gilbert Levin, who had designed The Labeled Release experiment and knew that its results were actually definitive.

The confusing results of the life experiments were explained by "weird chemistry" in the soil. A dozen Byzantine but lifeless chemical processes were trotted forth to explain the positive life results. The fact that many of these processes were shown to be inoperative as quickly as they were proposed was not considered important; everyone knew Mars was like the Moon. Besides, did not Ockham's Razor say that the "simplest hypothesis" was always the correct one? Was not life complex, and therefore never the simplest hypothesis? So the life experiment results, and every other strange thing seen on Mars, were automatically given simple, and therefore, non-biological explanations. This reflex rejection of "life-as-never-the-simplest-hypothesis" became ingrained in Mars science. Just about any hypothesis, no matter how arcane, will be entertained at a Mars conference, as long as it does not involve biology or the conditions conducive to it. This mindset has continued since the Viking life experiments in 1976 and has led to a crisis in science.

Back on Mars, however, the rest of the Viking lander team was having a happier time. The determination of the atmospheric composition were quite fascinating and so were the mineralogical and chemical analysis of the soil. The atmosphere was mostly carbon dioxide, but it had a few percent nitrogen, considered necessary for plant life to live. Beyond that it was one percent argon, a generally harmless gas similar to helium, and found in the Earth's atmosphere. However, a strange thing was found the fourth most abundant gas in Mars atmosphere was oxygen, not a large amount, an eighth of a percent, but much more than expected. Carbon monoxide,was found also, but only half as much as the oxygen.

It had been expected that carbon monoxide would be present in much larger amounts than oxygen, since it was formed by ultraviolet light striking the carbon dioxide molecules and splitting off an oxygen molecule. This expected process would make two molecules of carbon monoxide for every oxygen molecule, but the relationship was reversed. Two times as much oxygen as carbon monoxide was found. Added to this the oxygen was released as atomic oxygen and was supposed to combine with the soil or float away into space leaving the carbon monoxide to dominate. Instead molecular oxygen was dominant. Somehow, the orphan oxygen atoms were finding each other and making molecular oxygen. It did not make sense. The soil had been found to contain oxygen also, and to release it when exposed to water or heat. This had been an enormous headache to the life experiment researchers. Water vapor in interesting amounts was also found, but given the cold and thin atmosphere the air was very dry by Earth standards. This raised the perplexing question of where the water vapor was coming from, and how much was there at the source, but the soil held distractions from further debates.

Viking lander instruments did a complete chemical assay of the soil.The results were surprising but in a pleasing way. To the relief of the lander team the results raised no immediate questions concerning biology. The soil was full of iron in its most oxidized state called hematite. The age-old question of why Mars was red was answered: It was full of rusty iron. The soil was otherwise unremarkable: To most humans, it looked like everyday dirt from some region with a lot of old volcanoes. Later a strong match would be made with volcanic soil from the slopes of Hawaiian volcanoes. Another good match was to mix 50-50 pulverized basalt with powered CI meteorite. CI meteorites were an exotic meteorite believed to be clumps of the primordial dust from the nebula from which the planets formed. Because of this apparent primordial origin, the CI composition was used as a standard to measure the composition of samples of anything from outer space. Their composition was thus well known. However, after some discussion, the idea of the CI or any other meteorite originating from Mars was dismissed by the survey team as completely implausible. They quickly moved on to other fascinating things found in the soil and other things fascinating because they were not found.

The early soil analysis produced an interesting oceanic clay called montmorillonite. It was also salty, and was encrusted with a sulfate rich "caliche," like Plaster of Paris, as found in dry lake beds. Lots of the soil particles were magnetic, suggesting headaches for future explorers.

However, no nitrates or carbonates were found, and this was unexpected. Nitrates were expected since nitrogen and water vapor were in the atmosphere and the ultraviolet light would break up the family of existing molecules so that, when molecules reformed randomly, some nitric acid would form. Nitric acid forms from hydrogen, oxygen, and nitrogen on Earth whenever a large amount of energy is injected into the atmosphere, such as by lightning, hot flames, or nuclear explosions. Plenty of nitric acid ingredients were found, and some scientists had predicted that nitrates would form a layer a meter thick on the surface because of the ultraviolet energy. Despite the fact that traces of nitric acid precursors were found in the atmosphere, no nitrates were detected. The biggest surprise was another absent mineral.

Despite the presence of water vapor and carbon dioxide in the atmosphere, which together make carbonic acid, no carbonates such as limestone were found in the soil. This was deeply puzzling since all ingredients for limestone or dolomite, calcium and magnesium, and carbonic acid were present, yet no carbonates were detected. Iron also forms carbonate under circumstances that should have been present on Mars, especially on an early wet warm Mars in the primordial chaos. But carbonates were missing in action.

The final mystery, so full of portent, was the isotopic composition of the atmosphere found by the landers. The elements are distinct chemically, but this is a property of the number of electrons they carry. Balancing the number of electrons to make the whole atom have zero charge are the same number of protons in the nucleus, thus the identity of the element is determined by its number of protons. This is called its atomic number. However the protons are all charged positive and yet are compressed together in a nucleus. Like charges repel so the whole thing should explode, and sometimes it does. What prevents the nucleus from exploding most of the time is a sort of nuclear glue carried by particles called neutrons, which true to their name, are neutral. The neutrons allow the protons to stick together in the nucleus, the nuclear glue is a short range force called the strong force, unimaginatively named because it is stronger than the electric force which causes like charges to repel. Elements can exist which have more or fewer neutrons than the standard number, resulting in the same chemical properties but different masses; atoms of the same element with different masses are called isotopes. Most elements have several stable isotopes.

Processes in nature can separate isotopes and these processes leave characteristic clues as to their nature. The simplest process is gravity, which allows lighter isotopes to rise to the top of the heap. Chemical reactions also sort isotopes by weight because the heavier isotopes move more slowly at the same temperature and so get left behind while the light isotopes react with other elements. Heavy water is formed by electrolysis of ordinary water, because the heavy isotope of hydrogen, deuterium, stays in the water while the light hydrogen reacts with the electricity and bubbles out as gas. This is why it takes large amounts of electricity to make heavy water. Similarly, uranium for nuclear work is enriched by sending it though filters as a gas;, the heavier uranium 238 is slower than uranium 235 that is used for nuclear fission; after passing though many filters, uranium hexafluoride gas becomes more and more dominated by U-235. This takes a massive operation and lots of power. In general when you see abnormal ratios of isotopes, it indicates powerful things have happened to alter their natural balance.

The isotopic ratios found in the Martian atmosphere were startling. The argon was almost all argon 40, a heavier argon isotope. This was much different than on Earth where argon 38 is present in large quantities. Argon 40 is formed by the decay of potassium 40, a long-lived radioactive element found naturally mixed with its stable isotope potassium 39. The preponderance of argon 40 suggested that the soil would be full of potassium; however, the potassium, normally a minor element in Earth soils, was barely detectable in the Mars soil. The oxygen isotopes were nearly Earthlike; however, the nitrogen was disturbed relative to Earth, and the heavier isotope was more abundant.In the heavy noble gas xenon, a bizarre thing was found. The xenon, instead of being spread evenly in abundance over its five different stable isotopes from 128 to 136, like on Earth or in meteorites, was almost all xenon 129. The isostope findings were astonishing.

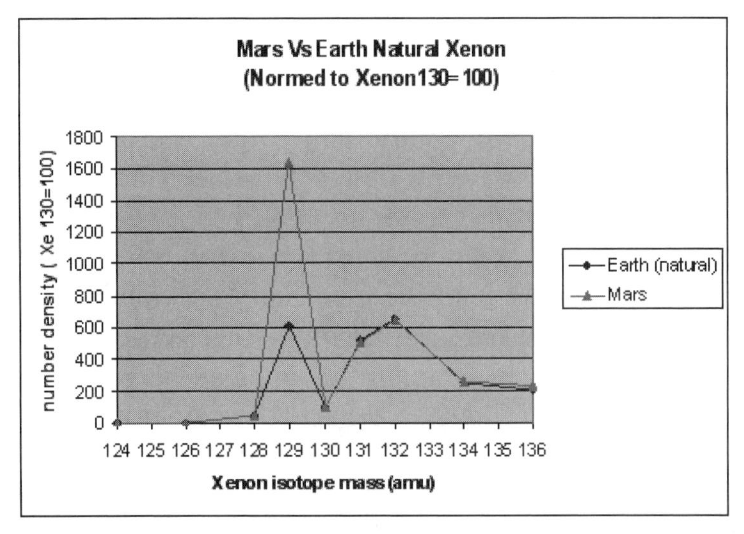

Mars and Earth atmospheric xenon isotopic abundance (JPL/NASA)

The second Viking arrived at Mars in early August after most of the shouting was over. Its mission was to support Viking 1 and was anticlimactic except for one nasty brawl over the landing site. The landing site was to be a place called Cydonia, but images from that area by Viking 1 showed it to be very rocky. One image in particular, from that imaging campaign caused a stir. It showed what looked like a large carved face, causing much controversy. The reader is referred to the excellent book *The Case for the Face*. The Viking 2 lander was finally set down in a place called Utopia, in what unexpectedly turned out to be a field of boulders, but which miraculously the lander had missed. The lander 2 experiments gave results strikingly similar to the Viking 2, including the life experiments and the soil analysis, despite landing on the other side of the planet at a much higher latitude. This similarity showed that forces on Mars had ensured the planet's surface was chemically very uniform.

All in all, the Viking probes were a glorious success. The skill, the mastery, and the daring and raw luck displayed by the whole Viking team continues to amaze. They handed to humanity the keys to an enormous library called Mars, full of mountains of information. The fact that some of the data was so surprising added in way to the glory; they were after all gatherers of data, and it was the glory of such data gatherers to find the unexpected. That they could not explain all of what they had found even added to the glory.

But at one point, it can now be surmised, men who were experts in nuclear physics and instrumentation, and who oversaw the isotope assay of the atmosphere, reacted when the xenon isotopic ratios in the atmosphere were announced, and stared across the conference table at each other with a momentary knowing glances. They then looked away lest any others would see their reaction. They had seen that xenon feature before, in their other work, in a place none dared to name.

Chapter 5. The Oxygen of Mars

"Reports of my death have been highly exaggerated."
Mark Twain

In the late 1970s, with Viking orbiters still operating around Mars and the landers still sending back data from its surface, a memorial service for Mars, of sorts, was held. The National Science Foundation reviewed the Viking life science experiments and declared that no life had been found. The search for Mars life was called off and Mars was declared dead. Some sang dirges of mourning for the old Mars. The distance humanity had traveled from 1900 in terms of Mars knowledge was immeasurable.

Gone long ago was the Mars of Lowell, Wells, and Dejah Thoris. The canals were long gone, so was the Mars that had lured humanity to launch the Mariners and Vikings. What had replaced it was a cold, battered world, with a wisp of an atmosphere and myriads of dried water channels that said that its best days were long past. The planetary community, for the most part, and the grant money, moved on to more interesting places.

The Voyagers were swinging past Jupiter in those days, dazzling the world with views of its Great Red Spot and flock of moons like a whole new solar system. On to Saturn and the icy kingdoms of the outer solar system the Voyagers plunged, their nuclear generators pouring out power and freeing them forever from any dependence on the Sun's dimming rays. The great planetary alignment was occurring, all the outer planets lining up in a row, and the Voyagers were taking advantage of it to achieve a "grand tour" of the major outer planets. So the yearners for Mars had solace in the exotic sights of Titan and Neptune. However, a strange thing had occurred. It was found that Mars, even though it had been declared dead, had been sending messages to Earth. The messages were etched in stone.

The Phoenix is a bird that dies in fire and then rises from its own ashes. In Japanese culture the Phoenix is known as the Vermilion bird (Suzaku), and associated with Mars. Like the Phoenix, Japan had risen from the ashes of World War II. By the later 1960s it had quietly resumed its place among the foremost scientific nations on Earth and established a research station in Antarctica near the Yamato Mountains. In 1969, a Japanese team in Antarctica discovered a trove of meteorites in the ice at the end of a glacier. Antarctica, it seemed, had been collecting meteorites falling from the sky on a moving conveyor belt of ice, a glacier, and then dropping them at the foot of a mountain range like a hamster storing seeds for the winter. Among these meteorites were many familiar types, for meteorites belong in families, grouped according to minerals and isotopes.

The Phoenix reborn

Some of the meteorites that had been found in Antarctica however, were of a very strange type, for they matched closely rocks brought back from the Moon. That is in fact where they were from. This demonstrated that meteorites could come from a large planet- like body, not just be blown off a small asteroid. This made people look with new eyes at a small grouping of meteorites that had perplexed experts for decades.

Collection of a meteorite in Antarctica (NASA)

Meteorites are named from the places where they have fallen. Accordingly, a meteorite that fell in Chassigny, France, aquired that name, so likewise did the fatal meteorite that fell in Nakhla, Egypt, and finally the one that fell near Shergotty, India. These three meteorites were thus called the SNC (Shergotty Nakhla Chassigny) meteorites. They not only had similar mineralogy, but similar isotopes.

One of the most powerful methods to classify meteorites is to analyze the isotopes of oxygen they contain. Oxygen is the third most abundant element in the universe, and it reacts with everything, thus finding a place in everything familiar. Oxygen is important to humans because they breathe it. Combined with hydrogen it forms water, which makes up 90 percent of our body weight. However, even the rocks we stand on are full of oxygen; the oxides of silicon and various metals such as iron, aluminum, magnesium, and calcium form the building blocks of most minerals. So oxygen is found in everything but it also marks everything with an isotopic identity tag.

Oxygen occurs in three stable isotopes, of atomic weights 16, 17, and 18 units, and the ratio of these among each other is like a mathematical fingerprint. Oxygen 16 is the most abundant isotope so it is used as a reference, the ratio of the abundance of oxygen 17 to the amount of 18 is what is key. Using oxygen isotopes, one can plot a map and place everything that falls out of the sky on it. Meteorites form tight clusters or lines on such a graph when the 17 to 18 ratio is plotted.

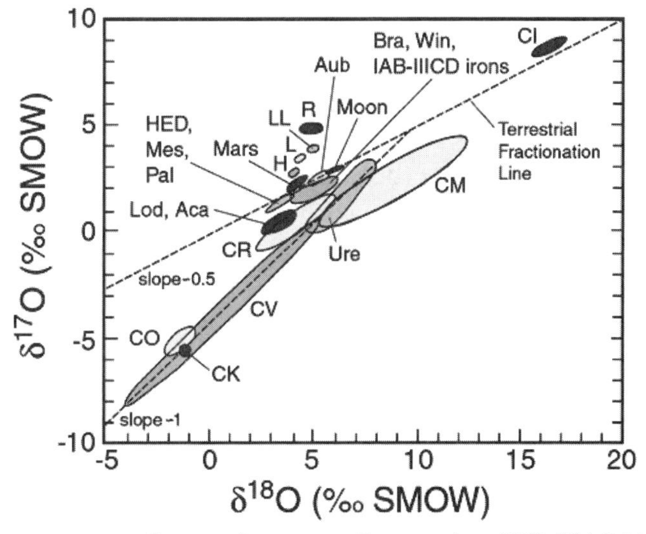

Oxygen isotopes of meteorites (JPL/NASA)

On this graph everything on Earth lies on a line running at a constant ratio of 1: 2.24 of Oxygen 17 to 18 abundance through the point 1, where 1 is the terrestrial fractionation line. Everything found in nature on Earth falls on this line, from the sea and land. Rocks from the Moon also fall on this line, confirming that the Moon was not captured as it roamed around the early solar system, but was made from the same isolated pot of molten rock as the Earth. The Moon rocks all sit at one point on the line because there is no liquid water on the moon. It is the interaction on a molecular level between liquid water and minerals that smears the oxygen coordinates of earthly rocks and minerals into a line rather than a point like the Moon rocks. Water contains oxygen like the rocks, and has the remarkable property that it "swaps" oxygen with minerals, and does so in a manner that favors the lighter and swifter oxygen 17 consistent with its mass difference. Thus, the slope of the line. Minerals with lots of interaction with water, weathered clays and water-deposited limestone, all have "lighter" oxygen dominated by oxygen 17 and thus lie higher up on the line from dry volcanic basalts. On Earth, the water and rock, the two great reservoirs of oxygen, share the same fractionation line because of plate tectonics, where surface rocks are pulled down into the bowels of the Earth with the waters of the ocean, crushed and heated together so that the oxygen atoms mix between them completely. They are then both, as steam and lava, shot back up to the surface from volcanic vents or summits with the same isotopic signature.

It is a glorious and mysterious thing to look at the oxygen isotope map of all things that fall from the heavens. The map is dominated by lines and clusters, most wrapped in mysteries dating from the foundation of the world. The terrestrial fractionation line dominates the map, another line runs nearly up and down the map. This is the CAM (calcium aluminum magnesium) line for the toughest, oldest pebbles in the solar system, recovered from the type of meteorites called chondrites, from Greek meaning "seeds' because of their welded-together pebble-like structures, which are impact shock-welded conglomerations of a grab bag of dust and minerals in the primordial chaos of the early solar system. Forming small regions on the graph above the terrestrial line are the L and H chondrites.

There are other clusters whose meaning is slowly emerging or remains enigmatic. Clusters or lines are thought to represent samples broken off a single "parent body" and carried to Earth after surfing various gravity fields, or even running into some small wandering chunk of rock. One group called the eucerites has been traced recently to the asteroid Vesta in the asteroid belt. Another mysterious group lies precisely on the terrestrial line and is called enstatites; however, they are from some utterly unearthly place. They look like jewel-studded crystal pulled from some furnace in the heavens. In all this mystery, the group of SNC (Shergotty, Nakhla and Chassigny) meteorites and their cousins, lay in a cluster just above the terrestrial line.

The people who study meteorites are called meteoriticists, a breed apart from planetologists, who are used to gathering information on what they can see every night in the sky but can never touch. Meteoriticists are people used to handling pieces of stone from unknown places and trying to figure out what those places are like from the rocks they have sent us. Unlike the planet people, who have flashy missions and tend to hold news conferences, the meteorite people tend to attend conferences and go to Antarctica to gather interesting rocks. The meteoritics people spend long hours staring through microscopes at thin slices of rock specimens. Planetologists study the big stars roaming the night; meteoritics people study streaks of light that are gone in a heartbeat. But all of that was about to change.

The SNC group had been noted, and grouped together by meteoriticists, for decades for being young lavas, ranging from 1.3 to 0.3 billion years old. These ages were determined by radio-isotopic methods well proven on Earth. Many meteorites look like lavas; however, they are all approximately 4.5 billion years old, and thus date from the formation of the solar system. Even the youngest lunar lava meteorites from the Moon are 3.2 billion years or older in age. From the SNCs it could be seen that some place in the solar system was throwing rocks at Earth, and that place had volcanoes that were recently active. Also, the place had to be close, because the meteoritics people could tell from cosmic ray effects that the meteorites had been flying around in orbit for only a few million years, a shorter time than many other meteorites that were believed to come from the asteroid belt. Somewhere, a planet-sized body with volcanoes, was blowing smaller rocks back into space, so they would orbit the Sun a few million years and end up falling on Earth. Venus seemed unlikely because it had powerful gravity like the Earth, a dense atmosphere that would make escape from its surface very difficult, and was closer to the Sun, making rocks ejected from it unlikely to move outward in orbit to Earth. With a combination of logic and astute observation that would impress Sherlock Holmes, some meteorite scientists in the late 1970s arrived at a startling hypothesis: The parent body of the SNC meteorites was Mars.

From Mariner 9 and Viking probes it had been learned that Mars had young volcanoes. This was evidence that the SNC were Martian but something more compelling was required to prove it. Fortune smiled on the meteoritic scientists in Antarctica, in 1979, in the newly found meteorite gathering fields. A meteorite was found at a place called Elephant Moraine. The meteorite was cataloged as ETA979001, ETA standing for Elephant Moraine, 79 for the year and 001 marking it the first meteorite of that year.

The meteorite turned out to be the "Rosetta stone" of Mars, for, trapped in the obsidian-like melted rock that formed part of it, were bubbles of gas. These gases were analyzed for isotopes in 1982 and showed the same remarkable xenon 129 excess and argon 40 excess that had been found by the Viking landers on the Martian surface. Soon other SNC-type meteorites were tested in this fashion. Finally, by 1985, a consensus formed, the young lavas with lots of argon 40 and xenon 129, were samples of Mars, blown clear to Earth by meteor impacts on Mars. It was a stunning and marvelous discovery. Nature had provided free of charge what planetologists had only dreamed of getting after an expensive and complicated space mission: samples of Mars.

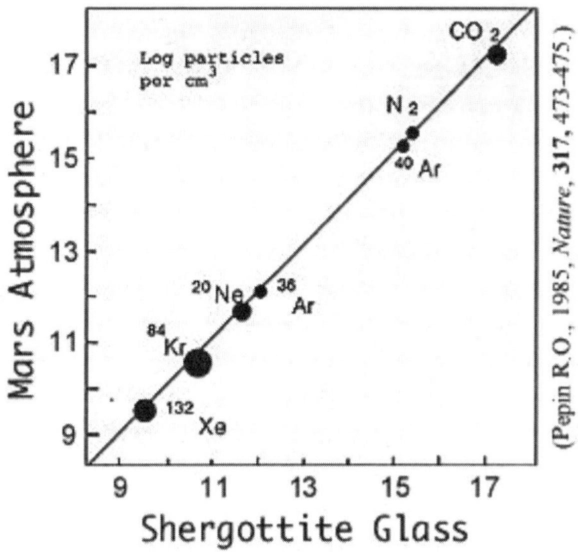

A graph of Mars atmosphere versus traped gasses in Mars meteorites

Mars, declared dead in 1979, was suddenly enticing and beckoning humanity again, five years later. It was not only showing signs of life, it was serving us appetizers. The free samples were tasty indeed. The minerology of the SNC group was neither simple nor predicable. They hinted at origins in a rich and highly evolved geologic system on Mars. The lavas that made up the SNC group were varied, ranging from iron-rich basalt similar to that from the volcanoes of Hawaii; these were the Shergottites. But the Nakahlites were rich in calcium, and Chassigny was nearly pure peridot, an emerald green semiprecious gem, known to geologists as olivine. Strangely, the rocks were low in potassium, uranium, and thorium compared to average Earth rocks, and they had substantially more iron. This contradicted the seemingly reasonable results of the Mars spacecraft sent by the Russians. More meteorites from Antarctica and from the Sahara were found that belonged in the SNC group. The buffet of free samples of Mars was increasingly varied and strange.

The SNCs had many unexpected properties. They were all geologically young, with radiometric ages ranging from 0.18 to 1.3 eons old. Their relative youth had connected them to Mars, but few scientists thought of the implications when making this connection. All of them were young, not just some of them, and this suggested that Mars was a deceiver, for it looked old, dead, and battered, but the SNC said that parts of it were young and dynamic. The rocks also had another intriguing property; they had water-deposited minerals in them, and these crystals of gypsum and carbonate had to be young also.

Groundwater had to be moving around on Mars, below the surface. The rocks had water trapped in them, a few drops per kilogram, and this water told of unearthly things. With slightly more oxygen 17, the water oxygen isotopes were lighter than the rock in which they were embedded. All the SNCs contained lighter water oxygen. This confirmed the Viking imagery which had shown no evidence that plate tectonics had operated on Mars like on Earth. So Mars rock and water had kept a separate oxygen signature. The waterborne minerals were much different in their oxygen isotopes from the rock. Also, they were smeared out from the rock on a fractionation line similar to the terrestrial line, but the line was above and parallel to the terrestrial line. Mars had its own line. This meant that like Earth, Mars had been a system of geology and water moving and interacting. The young age meant that this had been true recently in geologic time, as recently as 180 million years ago, at the time when

Jurassic age dinosaurs, like the stegosaurus and the allosaurus had ruled the Earth.

The SNCs told further of oxygen on Mars. Substantial iron minerals in them were not the normal ferrous or low oxygen state, which is black like basalt, but in the high oxygen state called ferric, which is red like hematite. The rest of the minerals, especially the waterborne minerals were all high oxygen states. It was as if the ground water of Mars had been on the surface, and been exposed to oxygen there, before sinking into the rocks. Earth's soil is highly oxidized;, it contains several atmospheres worth of oxygen, if it were heated and the oxygen driven back into the air. The same can be said of Mars. The Vallis Marineris exposes sediments 8 km thick, and they are all light red colored and therefore highly oxidized. It has been argued by some that the walls are simply dusted with windblown red dust, but the fact that dark low oxidation state lava deposits can be seen on the Vallis floor shows this is not the case, for they are not covered with dust and they are lying flat, while the walls are vertical. Therefore, Mars is oxidized in depth, not just on the surface. This means that the surface conditions were oxidizing when the sedimentary layers were laid down eons ago. Mars is highly oxidized all over its surface also, with only a few dark areas of lower oxidation showing.

This was strangely in agreement with the Viking results, where not only was the soil bright red and full of hematite, but it even exhaled oxygen when water was applied to it. These things were seen in the rocks by meteoriticists and recognized because they were familiar; the Earth had rocks with similarly highly oxidized minerals. But the meteoriticists shook their heads quickly: Such comparisons were dangerous and probably misleading, for Earth had an oxygen atmosphere. It was probably misleading, because the rocks were from another planet, whose geologic history was unknown, and it was dangerous because it suggested life.

Meteoritics had suddenly been rocketed to center stage, on Mars, that most treacherous of worlds. The meteoricists, being by nature cautious people fond of rocks, did not want to do anything to spoil their moment of fame.

To be a meteorite specialist had become cool; they were now being invited to parties. They had become planetologists, and planetologists of no mean parent body, but the most fascinating planet of all Mars. Like any sample of something that has only been observed remotely, the SNCs were both confirming and overturning ideas about Mars from the Vikings. The impact of the SNC on Mars studies was electric and unfolding. It contributed strongly to a Renaissance of Mars in the mid 80s that gathered steam as the Cold War was ending. Not only did it suggest that Mars was a dynamic and water rich planet in the past, as had been found by the Vikings, but it suggested that this activity and the planet's surface itself were both recent. But the most telling shock was yet to come.

The SNCs contained other things besides minerals: Mars was showing life signs. In the minerals of the SNCs were traces of organic matter, the stuff of life. The meteorite where these were found in largest concentration was ETA79001, the Rosetta stone of Mars and also one of its youngest. The Nakhla meteorite also contained substantial organic matter. The verdict of a lifeless Mars during the Viking missions had been based on the absence of detectable organic matter on the Mars surface. This report from the meteoritics people made that verdict look hasty and hollow. The organic matter was found in highest concentrations in a vein of water-deposited minerals found in the center of ETA79001.

Organic matter is part of the primordial stuff of the solar system. It is found in other groups of meteorites called carbonaceous chondrites. These meteorites are considered very primitive remnants made of the dust that formed the planets. They are very ancient, 4.5 eons old. Because of them, it was commonly believed that organic matter was part of the original ingredients of Earth, Mars, and the other planets in our solar system. But this ancient organic matter did not mean life; it was simply its as yet dead building blocks. Such organic matter would have been oxidized and burned up by ultraviolet energy in the Mars environment long ago. Therefore, finding organic matter in a meteorite was not considered startling, but finding it in a meteorite from Mars was, especially one that was young.

The Mars enigma deepened as more meteorites were gathered and a steady stream of new SNC types were found. The meteorites were still bouncing in ages between 180 million years in age and 1.3 billion years, a younger range than expected. The meteorites came from multiple impacts on Mars, known because their space exposure ages from cosmic ray tracks were in several clusters. This result, combined with the young ages, suggested that a large part of the Martian surface was younger than it looked based on crater counting. However, the deepest problem was the absence of any older ages. Mars had a dichotomy: that the north hemisphere was young could be accepted, but the south was perhaps 4.5 billion years old. Why were no southern meteorites being found? A partial answer came quickly.

In 1984, in Antarctica, a team looking for meteorites spotted a greenish lava looking rock lying in the snow., The team picked up the intriguing rock and later found it was indeed a meteorite and gave it the name ALH84001.It looked less interesting back in the lab than in the snow, and was classified as a diogenite, a curious but obscure group of lava meteorites, and put on a shelf. However, a worker named Middelfelhdt, looking through the old meteorite collections, found its greenish color intriguing. He analyzed it carefully, and based on years of experience, realized it had been misclassified as a diogenite. Based on chemistry and mineralogy alone, he boldly proclaimed it was Martian and belonged with the SNCs. This was quickly confirmed by oxygen isotopes. The real shock came quickly: its age was 4.6 billion years old. An old Mars meteorite, perhaps from the ancient southern part of Mars, had finally been found. A frenzied check of old lava meteorite collections was made, with the hope that the misclassification of ALH84001 meant that other ancient Mars meteorites had been misidentified. Alas, the misclassification of ALH84001 was apparently a fluke.

ALH84001 was a sample of primordial Mars, apparently the only one in collections. Because, at that primordial time Mars and Earth were assumed to have been similar in surface conditions and atmosphere, it seemed only reasonable to see if Mars had held biology as Earth apparently had held it. The results of this investigation stunned the world, for indeed, it appeared that the stuff of primitive life was preserved in the core of ALH84001, in a seam of carbonate rock apparently deposited by water.The sense that this was a forbidden result soon settled in, and an angry mob of scientists attacked the ALH84001 biology results. As one scientist from the ALH84001 life-group told the author: "we understand their criticisms, but not their anger." The ALH84001 scientists responded with evidence for life found in other Mars meteorites, such as Nakhla, but it did no good.

The old mindset of life-never-being-the-simplest-hypothesis awoke in full fury. A dozen arcane non-biological scenarios were invoked to explain the collection of chemical and morphological features found in ALH84001. As soon as one such non-biological mechanism was swatted down another dozen took its place. All of these non-biological scenarios were complex and some outrageously so, but all were entertained, even applauded, because they were all simpler than the simplest living cell.

Some scientists recalled a previous report of microfossils seen in a group of meteorites, called CI carbonaceous chondrites, that had been shouted down in the 1960s. These life-signs had been dismissed as terrestrial contamination. Those who recalled this affair, said that any meteorite that fell to Earth was by definition contaminated by the biology of Earth, and could not be used to prove anything about extraterrestrial life. One would have to go to Mars, and bring back rocks in a sterile container, it was said, to ever prove the proposition that life had existed on Mars. The planetary community finally shrugged its shoulders, and said that the life evidence in ALH84001 and the other Mars meteorites, such as Nakhla, was "not compelling."

While the life debate raged, the collection of Mars meteorites continued to grow and enrich itself, forming its own subgroups of age and mineralogy; however, ALH84001 remained the only ancient Mars meteorite. The range of ages between theShergottites, which tended to be 180 million years old, and the Nakhlites and Chassigny, which were 1.3 billion years old, was filling in with the discovery of meteorites of intermediate age. However, the "age paradox" deepened.

The Mars Age Paradox was two-fold: 1. The group of young meteorites was younger than expected even if all came from the young half of the Mars dichotomy. 2. Half the expected Mars meteorite collection was missing, the old half from the old part of the Mars dichotomy. The fact that one ancient meteorite had been found from Mars made the mystery even deeper. The statistics were totally wrong for young and old contributions to the collections. The statistics suggested that large areas of Mars were young, and that the obviously old areas of Mars were not producing recognizable meteorites.

ETA79001 The Rosetta Stone of Mars, the first proven Mars meteorite (NASA)

A solution to the first part of the paradox was found. The difference between the ages of the meteorites and ages of the surfaces they came from was found to depend on one unknown parameter, the cratering rate on Mars. The meteorite ages could be measured directly and were assumed to be rock from a layer near the surface that was "sampled" by being sent into space when a large asteroid impacted Mars. However, the surface age of regions of Mars was an estimate, based on an estimated rate of meteorite bombardment on Mars compared to the Moon. This was the unknown parameter. It had been proposed that the cratering flux on Mars was between 1x lunar and 4x lunar, with 2x lunar being the best estimate. So the cratering flux on Mars was assumed to be enhanced by Mars's proximity to the asteroid belt. It was proposed by the author in 1996 that the meteorite flux onMars was actually much higher than estimated, at least 4x lunar. This suggestion was also made by Nyquist, in 1998.

If the meteorite flux was 4x lunar or higher, the ages of much of Mars's northern regions became younger, and the regions between 1.3 billion and 180 million years old became a much larger percentage of the northern area of Mars. This idea had also been proposed by Nyquist. So the first part of the age paradox could be solved by adjusting the crater rates on Mars. Such a high cratering rate would also explain why so much more Mars material, kilograms, has been recovered on Earth compared to the relatively small amount of material, a kilogram or less, than has been recovered from the Moon meteorites. Given the great distance of Mars and higher velocity to which fragments must be accelerated to escape Mars gravity in order to reach Earth, as opposed to those from the Moon, the cratering rate on Mars must be much larger, perhaps an order of magnitude, in order to create this excess of Martian material. Mars is apparently being hammered by the nearby asteroid belt.

The second part of the age paradox, the missing old meteorites of Mars, had an even more startling solution that remains controversial. The meteorites from Mars, including ALH84001 and all the SNCs, had been lavas. The frantic searches for missing Mars meteorites in the collections had concentrated on lavas. The only other types of meteorites were chondrites, and it was thought that they couldn't be from Mars. The chondrites were the right age, 4.5 billion years, but they were made of space debris of all sizes, shock-welded together by impacts. As strange as Mars was, it was not outer space; it was a planet with an atmosphere that slowed things down, especially small things. But meteorites had always been classified as chondrites, or achondrites. Achondrites were melted, semi-homogenous rocks like lavas, chondrites were completely heterogeneous mixtures of rocks that had never melted together and mixed. The puzzlement in the meteoritics community deepened with the discovery of every new young lava from Mars.

ALH84001, still the only ancient meteorite of Mars, holds signs of life (JPL/NASA)

That the missing meteorites were already in the meteorite collections seemed highly likely. They were most certainly hiding there, because the process of ejection of Mars rocks seemed to be flinging rocks to space from Mars, and hence to Earth, without shattering or melting many of them in the process. ETA79001 had had a rough ride from Mars, and was partly melted because of it. Nahkla, however, looked as if it had ridden here on a silk cushion. No evidence of shocking or melting could be found in Nakhla. This meant, among other things, that the process that flung rocks into space from Mars could bring a wide variety of rocks here. No one could think of a good reason that old rocks from Mars should not be sent here as well as young ones. ALH84001 had showed that old rocks from Mars seemed to get here by the same process as new ones.The question was then asked, what would a piece of the old surface of Mars look like? Would it be a lava, or perhaps a sedimentary rock?

The southern highlands of Mars were covered with old water channels, and the process of erosion had wiped out all craters smaller than 30 kilometers in diameter. That is a lot of erosion. Such erosion would create clay and mud, and the mud would be on the bottom of lakes that would form in craters, and then the mud would harden when the water dried up. The rock would look like shale or adobe. Could such rock survive being blasted into space from Mars? Was it possible that the missing meteorites of Mars were not lavas, but something else, perhaps something that looked like old lake bottom? The answer to this puzzle lay on the oxygen isotope maps of meteorites. The answer was so stunning to meteorite specialists that they initially recoiled in astonishment. The answer was forbidden.

The proposal by the author in 1996 that so stunned the meteoritics community was that the missing old meteorites of Mars were sitting in glass cases in museums, a few feet from the other Mars meteorites. These were the same meteorites that had caused a bitter life-debate in the 1960s.

The Orgiuel Meteorite

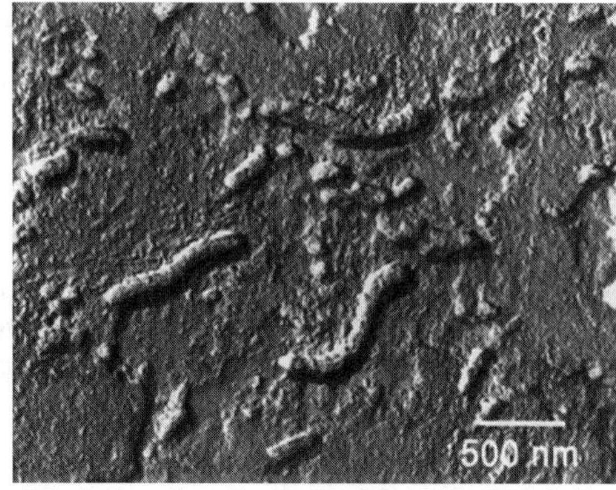

Possible Microfossils in ALH84001 (NASA)

They were a rare type called CI carbonaceous chondrites, and they consisted not of lava, but of hydrated clay. They were in a sense achondrites, because they have no chondrules, but classified as chondrites nonetheless because they were not lavas. The isotopes of oxygen and other gases matched Mars well, and they were 4.5 billion years old, and there were a dozen of them, about the same number as the SNCs. Therefore, they indeed looked like they could be the missing old meteorites of Mars. The suggestion they were Martian ignited a firestorm. It did not matter that the CI shared numerous properties with the other Mars meteorites, one thing made them unacceptable as Martian: They were full of organic matter and microfossils. If the CI were Martian, then Mars had been alive, and this was unacceptable.

Chapter 6. The Paleo-Ocean of Mars

"Follow the Water" NASA Official

While the discovery of life signs in a Mars rock occurred after a decade of discoveries concerning Mars meteorites, the rest of the Mars science community was going through its own revolution in its conception of Mars. The lunar Mars model was losing ground to a more terrestrial Mars model. While everyone had been watching the Mars meteorite show, the concept of Mars had undergone a sea change. Mars had acquired a paleo-ocean on its northern plains.

The paleo-ocean filling the northern plains was proposed by the author in 1986 at the Washington, DC MECA meeting. This hypothesis was based on widely recognized features of Mars data. It had been proposed earlier by David Chandler in his book, Life on Mars (1980). The Mars community, however, has insisted on giving credit to Timothy Parker, despite the fact that he proposed only small seas. After all, he was member of the tribe, and I was not. However, as its original proposer I have named it the Malacandrian Ocean, after the name of Mars in C.S. Lewis's *Perelandra* Trilogy. My involvement in the Cydonia debate had forever marked me on Mars as a dangerous person. Parker does deserve credit for bravery, for at the time, any such proposal was considered radical, and it must be known that science punishes radical views and those who proclaim them. They are denied tenure, and even funding in some cases. However, happily in this case, Parker survived and much good work by Parker and Head has buttressed the evidence for the northern paleo-ocean. The paleo-ocean appears to have had several sizes during Mars history, and appears to have shrunk in time.

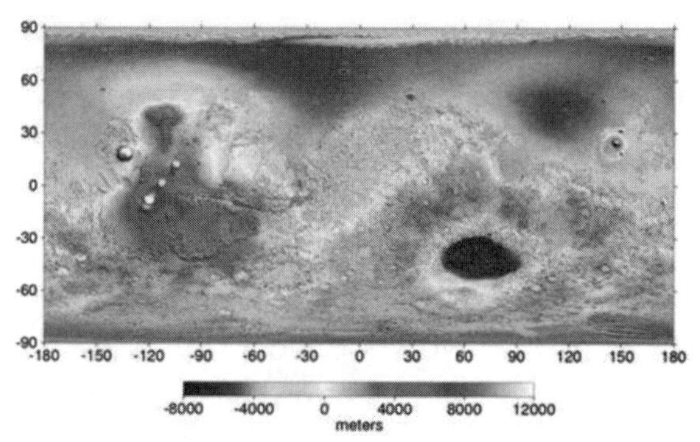

Mars's northern plains form an ocean basin (JPL/NASA)

The hypothesis that Mars had a paleo-ocean of large extent in area and duration has been motivated by observations of geomorphology, the study of landforms, more than anything else. The northern plains region appears to be an old ocean basin, and the numerous water channels appear to have emptied into it. On a smaller scale, several craters in the southern highlands appear to have held lakes. The northern paleo-ocean was the destination for countless Martian rivers that carved channels or vast areas of the Martian north.

The second strong reason for an ocean filling the Martian plains was estimates of how much water had flowed into all the river channels of Mars. The old river beds of Mars are large, and stretch for hundreds of kilometers. In some places it is apparent that catastrophic floods have been unleashed, sweeping all before them to the ocean shoreline. The flood waters appeared to have flowed for months, based on the eroded islands they left. When the amount of water flowing in these channels was estimated, it yielded the equivalent of 400 meters deep layer covering the entire planet. The ocean basin in the north covered approximately one-fourth of the planet's surface and had an average depth below "Martian sea level" of 1.5 kilometers. Thus the 400 meter deep layer was a good match.

Delta formed by long flowing water on Mars (JPL/NASA)

The fact that the edge of the basin all around its circumference, a radical change in landform from rugged and crater-pocked to smooth and much less cratered at a common elevation, presciently termed "Martian sea level." This term had been coined earlier by Mars cartographers and was based on the fact that barometric pressure, 6 millibars at that elevation and below it, allowed liquid water to exist. This coincidence with a change in landform solidified the interpretation that it represented the edge of an ocean basin. Thus even on modern Mars, enough atmosphere existed to keep water liquid in the ocean basin at temperatures just above freezing. However, there was other evidence for existence of the ocean.

The Viking landers had both landed on the smooth plains of the old ocean basin, and the soil they analyzed at each site was remarkably similar, and equally salty. This showed that some force had been at work to ensure remarkable chemical equilibrium, despite the fact that the Viking landing sites were separated by thousands of kilometers. This was particularly true of water soluble salts of chlorine and bromine, as are found in terrestrial sea salt. Added to this, the soil composition resembled oceanic clays on Earth.

The Vikings and Pathfinder landed on the Paleo-Ocean bottom (JPL/NASA)

The isotopes of Mars hydrogen suggested that much water had been lost over Mars's history. Mars has no oxygen or ozone layer to protect the surface from ultraviolet light. The ultraviolet light on Mars is so harsh it gives even water molecules a "sunburn," breaking them down into hydrogen and oxygen. The oxygen and hydrogen, both being lighter than the carbon dioxide atmosphere, rise, and in the case of hydrogen, are lost to space. However, the hydrogen is not all the same and so heavier hydrogen isotopes escape more slowly than ordinary hydrogen. Hydrogen in nature comes in two varieties: ordinary hydrogen with a simple nucleus of one proton, and at approximately 1 percent, a heavy hydrogen isotope with a nucleus of one proton and one neutron. It is this form of hydrogen, purified and concentrated and combined chemically with oxygen, which was the infamous heavy water used in the Nazi atomic bomb effort in World War II. The heavy hydrogen is twice the weight of ordinary hydrogen and thus does not float as high statistically in the atmosphere when it rises. The lighter hydrogen floats higher on average and escapes to space via a number of processes, such as a sandblasting effect of the impact of solar wind particles on the top of Mars's atmosphere. Therefore, the hydrogen in the Mars atmosphere becomes heavier in average isotopic weight over the eons as the process continues. This means the degree of fractionation of isotopes, the sorting of the isotopes by weight, increases with geologic time. Earth, with its ozone layer and heavier gravity, and magnetic field that prevents the solar wind from affecting the top of the atmosphere, prevents fractionation to the degree that would occur on Mars. Therefore, adopting the Earth ratio of hydrogen isotopic abundance as a yardstick, we can measure the different ratio of abundance of isotopes on Mars and use this as an indicator of the history of Mars.

Based on Earth and Solar hydrogen standards, the isotopes of hydrogen and other light elements are highly fractionated on Mars. This indicates that Mars has suffered much atmospheric loss by processes in the upper atmosphere. Hydrogen fractionates easily, so its large fractionation was not a surprise. However, by comparing the amount of fractionation of nitrogen, oxygen, and hydrogen on Mars a marvelous thing was discovered. It was found that the oxygen was hardly fractioned at all relative to hydrogen and nitrogen. When the models were run, this relative fractionation indicated that a vast reservoir of water had existed and had replenished the oxygen in the atmosphere even as water molecules were broken up and the hydrogen and oxygen were lost to space. The amount of water was estimated to be approximately enough to cover Mars's surface to the depth of 300 meters, almost the same amount estimated by the geologists looking at the flood plains. So by two widely different methods approximately enough water to cover Mars to the depth of ½ kilometer over its entire surface was found to have existed on Mars in the past. That was enough water to fill an ocean covering the northern plains of Mars.

Despite the geologic and nuclear isotope experts' opinions, the Mars climatic and atmospheric people have been hostile to the idea of a Mars ocean. The northern paleo-ocean of Mars, simply by its existence, has profound implications for Mars's geochemical history. If it had existed, it would have had an enormous effect on climate.

The fact that an ocean sat in the north of Mars means that the conditions on Mars had to have been Earth-like. The liquid state of water is one of the most demanding in terms of conditions of pressure and temperature. The presence of liquid oceans on Earth, defines terrestrial conditions of temperature and atmospheric pressure. Moreover, the presence of an ocean does not just define atmospheric conditions of pressure and temperature, it also stabilizes them.

The oceans of Earth act as an enormous "flywheel" on the atmosphere and climate of Earth. By flywheel we refer to the idea of mass in an engine that keeps it spinning at a nearly constant rate despite fluctuations in power and workload. Water has one of the largest heat capacities of any common substance. This means that when a quantity in a system is at a certain temperature, it will tend to maintain that temperature despite heat entering or leaving the system. If one adds energy, as when the Sun rises at, dawn, the water literally soaks it up and stays at nearly the same temperature. When the Sun sets, the water stays warm. This effect accounts for why regions of land near the ocean have milder climates than do regions inland, provided the prevailing winds are from the sea. Water also evaporates and condenses, and dissolves gases, thus stabilizing the atmosphere.

An ocean stabilizes atmospheric pressure because it can change from liquid to gas, but also because it can carry gas in solution. If the pressure of the atmosphere drops suddenly over an ocean, the water evaporates more rapidly and restores the gas pressure to a degree, if the pressure increases, what causes condensation to occur, lowering the pressure. The ocean water also carries dissolved gases. On Mars the dominant gas in solution would be carbon dioxide.

Carbon dioxide is very soluble in, water, with approximately each volume of water holding an equal volume of gas at STP: Standard Pressure and Temperature, or 1 atmosphere at 0 degrees Celsius. Anyone who has shaken a bottle of soda pop and then popped the lid has experienced this. On the surface of an ocean several kilometers deep with an Earth atmosphere of carbon dioxide above it, approximately an entire atmosphere could be held in solution, with much of it at lower depths where higher pressures allow more gas to be in solution. On Mars the ocean would be highly charged with carbon dioxide and if a sudden fluctuation in atmospheric pressure occurred, due perhaps to an asteroid impact, then the ocean would have frothed like a bottle of soda pop and the gas released would have made up for any pressure drop.

Thus, a Mars with a paleo-ocean would not only have had a climate like Earth, but the climate would have been stable. However, a specter haunts all studies of Mars, and that is the specter of Wells. This specter means that any new interpretation of Mars history that makes it more Earthlike is resisted mightily. This is because a paleo-ocean on Mars means the red planet held life, and a paleo-ocean on the northern plains means it held life for a long time.

Mars with the Malacandrian Ocean (JPL/NASA)

The ocean was believed to have been the original incubator of life on Earth, so the presence of an ocean on Mars is a powerful argument that life had begun on Mars. The watery environment supports and protects life from ultraviolet light andprovides a bath of minerals for growth. If Mars had an ocean, then living things probably lived in it. The question would be not whether life had begun on Mars, but how long the, ocean, and the life in it, had endured. With the discovery of the paleo-ocean, the answer, in the context of previous theories about Mars climate history, was shocking.

The Mars paleo-ocean was on the wrong side of the Martian dichotomy to be accepted. The ocean was on the youngest terrain of Mars, not the ancient part. It sat on the portions with few craters. Few would argue that an ocean could have temporarily existed in the early parts of Mars history, when many water channels were being formed in the southern highlands. However, the ocean did not sit in the middle of the ancient southern highlands; it sat in the north, on the low smooth plains surrounding the water ice cap. This seemed unacceptable. This would mean that the ocean would have existed on Mars, not for a brief period in the beginning, but for most of Mars's planetary history. Given the vagaries of Martian terrain dating schemes, the ocean may have lasted 4.0 billion years. The openly articulated reasons for why this seemed unacceptable had to do with the understanding of the planetary greenhouse.

Mars is too far from the Sun to allow Earthlike temperatures; it requires a boost to trap the heat of the sunlight. This effect occurs on both Earth and Venus because of a carbon dioxide greenhouse effect. This greenhouse effect of carbon dioxide is so strong that the surface temperature on Venus is higher than that on Mercury, even though Mercury is twice as close to the Sun. But on a planet like Mars, it was believed no greenhouse could have lasted for more than a few million years, certainly not 4 billion.

Chapter 7. The Crystal Palace of Mars

"Those who live in glass houses should not throw stones"
Old Saying

From the moment water channels were found on the Mars surface, suggesting a past warmer and denser atmosphere, the possibility of a planetary greenhouse appeared in the minds of researchers.

Carl Sagan pioneered the idea of a Mars greenhouse and its hazards

Carl Sagan had championed the idea of greenhouse effects on planets. He first proposed in the early 1960s that high temperatures would be found on Venus, and argued persuasively that this would be due to the greenhouse effect of its dense carbon dioxide atmosphere. These high temperatures were then observed, making him a hero in planetary circles. Venus, despite being twice as far from the Sun as planet Mercury, has a higher surface temperature. Venus had lived up to her ancient reputation as a hottie.

The transfer of energy from the Sun to a planet occurs when visible light from the Sun carries energy through an atmosphere transparent to visible light, to a planet's surface, where it turns into heat. All planets receive energy through electromagnetic radiation from the Sun and give it back into space by radiating electromagnetic energy at longer wavelengths of infrared, also called heat rays. That is the way energy is transferred from one body to another through the vacuum of space. The full electromagnetic spectrum ranges from radio waves with wavelengths of kilometers, to microwaves with wavelengths of centimeters, to infrared, or heat rays, with wavelengths from a hundredth of a millimeter to a thousandth. Infrared cannot be seen but can be felt on the hand and face by the skin.

Shorter still in wavelength is visible light. The rainbow of colors we see is only a small part of the electromagnetic spectrum, and ranges from, red, the longest, to violet, the shortest wavelength. Shorter than visible is ultraviolet which can be detected by the skin when it gets sunburnt. Basically the photons of ultraviolet are so full of energy they break up the protein molecules in the skin. Shorter still than ultraviolet are x-rays and finally deadly gamma rays are shortest and most powerful of all.

Basically the visible spectrum is a range of wavelengths that pass easily though water, which fills our eyeballs. However, in the infrared or in the shorter wavelengths of

Morning Frost on on the Paleo-Ocean bed of Mars (NASA)

Earth and Mars Shown to Scale (NASA)

Our Oasis in Space and its desolate Moon (NASA)

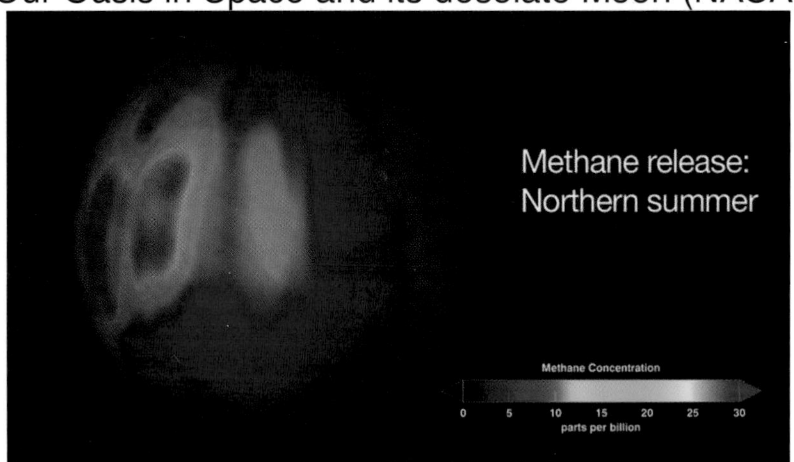

The Methane Plumes of Mars(NASA)

The Twin peaks of Mars. (NASA)

Earth as a Red Planet, note the redness of desert regions due to oxygen (NASA)

Mars, viewing Mare Acidalium, the dark area in the north, where whatever last catastrophe happened on Mars, happened.(NASA)

The Surface of Mars: Viking lands in Utopia Planum on the old ocean bed.(NASA)

Sunset on Mars , note blue sky. (NASA)

Victoria Crater on Mars viewed from a Mars rover. Note exposed water deposited sedimentary layers. Dark soil is Hematite. (NASA)

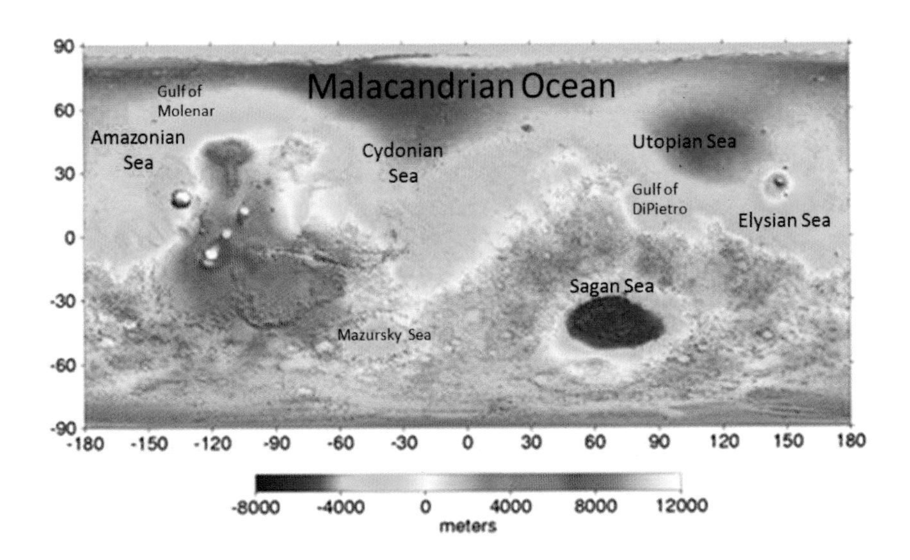

The Paleo-Ocean of Mars -elevation map, showing the old ocean basin in blue. (NASA)

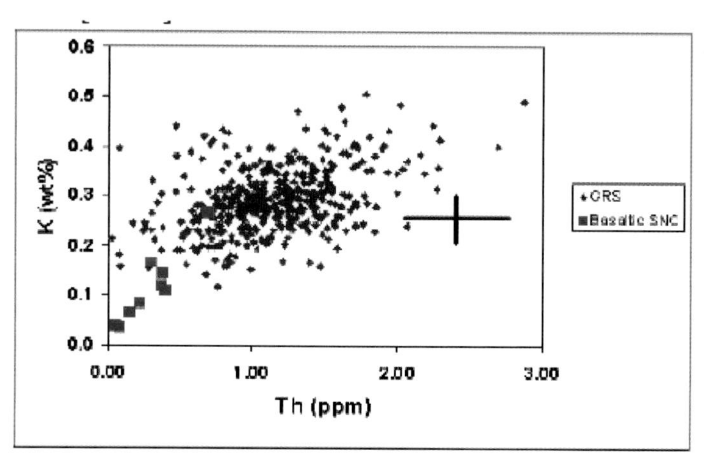

K, Th variations on Mars compared to Martian (SNC) basaltic meteorites. Typical statistical uncertainty shown on right.

Mars surface has much more radioactive Potassium and Thorium (GRS) than its meteorites (NASA)

Frozen Lake in a Mars crater. (NASA)

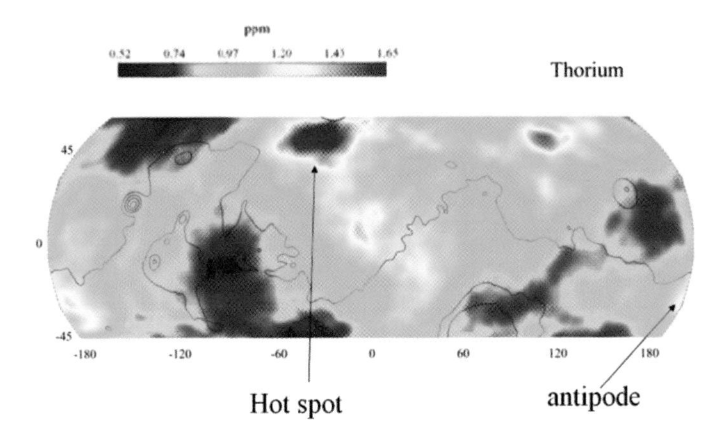

The Thorium of Mars, note the hot spot at the antipode.

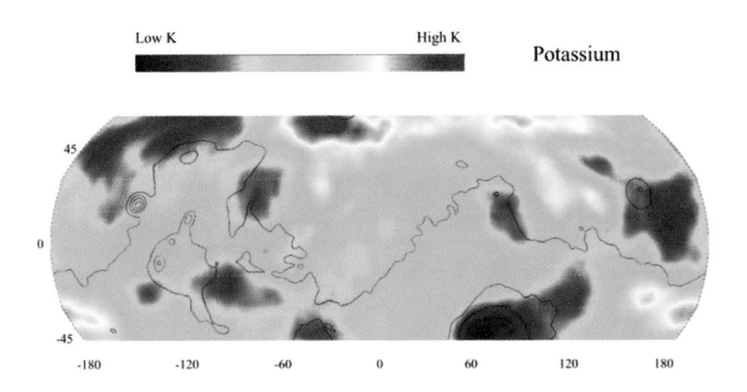

Map of radioactive Potassium on Mars, note the same distribution as of Thorium including the antipode.

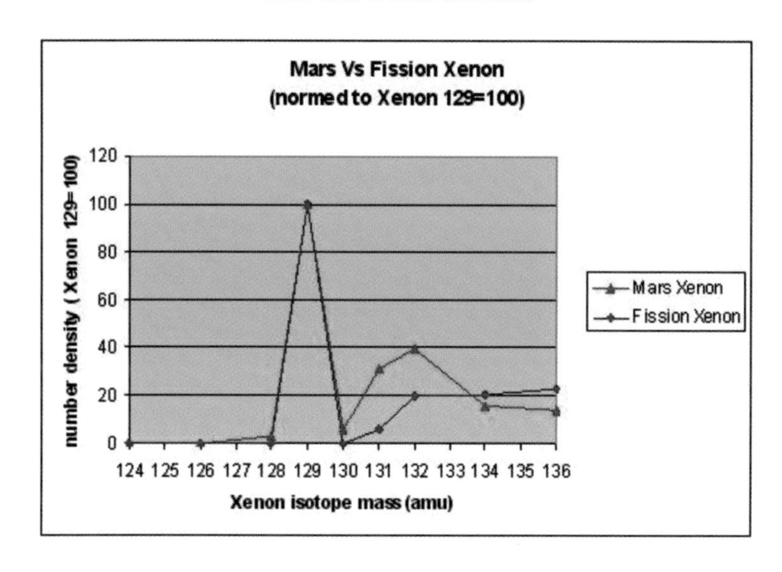

Mars and Human Fission produced Xenon Isotopes (data source JPL/NASA)

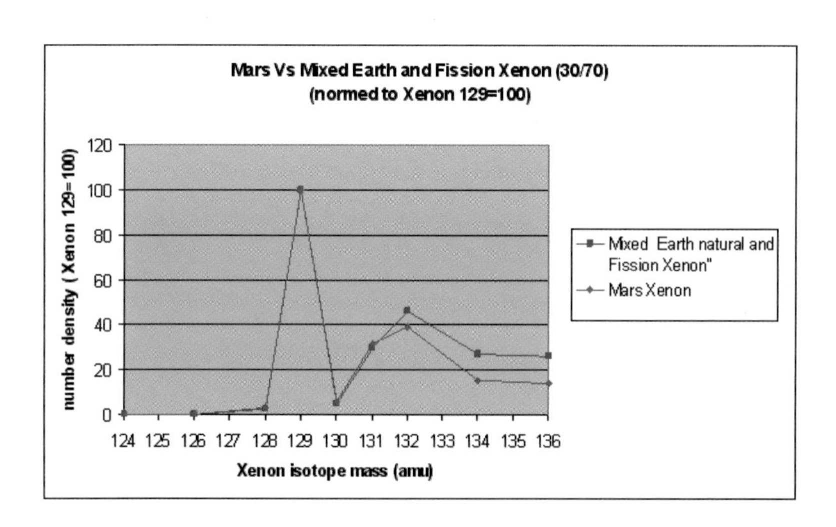

Mars and mixed Human Fission produced Xenon and natural Earth Xenon Isotopes

ultraviolet water is black as ink. Visible light is where long-lived stable stars like the Sun put out most of their energy over their long lives. The transparency of water in the visible, where so much other activity takes place, such as photosynthesis, is one of the miraculous coincidences of nature, since water is also vital for life.

By the process of radiation absorption, chiefly in the visible, and re-radiation of heat, chiefly in the infrared, the planets achieve radiative energy balance, and thus a steady temperature range. All hot things radiate energy in proportion to their temperature, and a spread of frequencies in the electromagnetic spectrum from radio waves to the visible. White-hot objects radiate energy primarily in the visible, like a light bulb filament or the Sun itself. Lukewarm objects radiate primarily in the infrared. The hot surface of the Sun being white hot radiates most of its energy in the visible rays, but the cooler surface of a planet radiates primarily infrared heat radiation.

The greenhouse effect occurs because the atmospheres of planets pass electromagnetic energy at different frequencies in different ways. The Earth has an atmosphere that passes visible wavelengths easily; the presence of much water in the atmosphere does not hinder this. The presence of water does block much radiation in the infrared and ultraviolet. It is for this reason that much of the Earth's surface is constantly viewed from space in the visible. Radio waves, at much longer wavelengths of electromagnetic energy, also travel long distances without difficulty whileultraviolet is absorbed by the ozone and so is blocked. In infrared of various wavelengths, water vapor, methane, and carbon dioxide absorb and re-emit the photons after a few hundred meters, which is why the Sidewinder missile, which uses heat rays, is only used for short-range air battles. At long ranges it is unreliable. For long-range air battles radar guided missiles are used. Infrared light that is blocked by the atmosphere must diffuse through it, with the infrared photons bouncing around randomly in the atmosphere like molecules rather than shining straight through like visible light. Carbon dioxide is not as effective at blocking infrared as methane or water, but anyone who doubts carbon dioxide molecules' affinity for infrared light should consider that carbon dioxide forms the working medium for infrared lasers used for cutting steel in industry, and they are the most efficient and rugged high powered infrared lasers known.

So if a planet has a dense carbon dioxide atmosphere the surface receives energy at visible wavelengths, but cannot cool itself efficiently by radiating it at long wavelengths. It is like the greenhouse design first built in the Victorian era, mostly glass, which passes visible light but reflects the infrared trying to make its way out. Therefore, the surface gets hotter and hotter until enough heat radiation can diffuse itself back up through the atmosphere to establish radiation balance. Venus shared a carbon dioxide atmosphere with Mars, albeit a much thicker one, and its greenhouse effect was very effective. Mars's present atmosphere is too thin for the carbon dioxide to give much greenhouse effect. However, Venus demonstrated that the greenhouse effect was a fact. Therefore, the simplest explanation for how Mars would have flowing rivers in the past seems to be that Mars began its existence with a much thicker carbon dioxide atmosphere like Venus, and that this trapped enough heat to melt water ice and simultaneously gave enough pressure to keep the water in a liquid state. Other gases such as methane and water vapor are even more effective, but water is mostly frozen on Mars, and methane is too fragile, being rapidly broken up by ultraviolet light without an ozone layer to protect it. Carbon dioxide is the greenhouse gas of choice because it is very rugged and remains a gas even below the freezing point of water.

Planetary modelers quickly found that a surface pressure of approximately 1 Earth atmosphere of pure carbon dioxide would raise the Mars surface to above freezing at present solar irradiance. Pressures higher than this would not be as effective, because at one atmosphere water was not only liquid but could also vaporize readily, and water is a much more effective greenhouse gas than carbon dioxide. Thus, at moderate pressures like Earth, a greenhouse of both carbon dioxide and water vapor would operate and allow an Earth-like Mars. However, it was shown quickly that this simple greenhouse model was surrounded by complex problems. Nothing was simple on Mars.

Naïve models of a Mars greenhouse as simply a layer of carbon, dioxide, as on Venus, failed on Mars for two reasons. The first is that Mars is farther from the Sun, and is inherently a cooler planet than Venus, thus carbon dioxide can condense to dry ice on the Mars surface at the poles. This problem was first addressed by the team of Sagan, Toon, and Geraish in 1972 who found that a Mars with a dense carbon dioxide atmosphere was unstable to catastrophic collapse.

Mars spins like Earth and has a polar tilt and polar caps like Earth. During the winters in the north and south the temperatures drop far below freezing, so low in fact that the carbon dioxide making up the bulk of the atmosphere actually freezes out partially as dry ice. During the rest of the year, some ten percent of the atmosphere solidifies on one polar cap or another. If the atmosphere on Mars were much denser, this effect would be much more, pronounced, since gas at high pressure condenses and solidifies more readily than gas at low pressure. The fact that at Earth-like atmospheric pressures, the carbon dioxide would trap a lot of heat in a greenhouse merely intensifies this instability. Once condensation starts at the poles, the atmospheric density drops as atmosphere freezes out, and the greenhouse effect depends on density, so the greenhouse effect would disappear, making it colder still, accelerating the condensation at the poles, until most of the atmosphere is frozen on the surface, and all of Mars is in deep freeze.

Mars with enough carbon dioxide to form an atmosphere of Earth-like pressure would be like a boulder poised on the edge of a high cliff. One big shove and the boulder falls to the bottom. The planet has two possible atmospheric states: one with a warm dense atmosphere, where much heat was trapped; the other with a cold thin atmosphere with most of its carbon dioxide frozen on the surface. The planet could run with a warm greenhouse but if a cold winter happened at one pole or another the planet could "fall off the cliff" into a "dry-ice age." We shall refer to this problem uncovered by Sagan and company as the "dry ice instability." If Mars froze, it was also possible for a warming event to pull Mars out of its dry-ice age back into a greenhouse mode; however, such a potential for catastrophic climate change, whipping from one mode to another, would make Mars unsuitable for most biology known on Earth.

The second reason a simple carbon dioxide greenhouse on Mars would fail quickly is the fact that carbon dioxide forms an acid, carbonic acid, in the presence of liquid water. This acid attacks the lava rocks found on Mars and forms carbonates and thus the carbon dioxide is removed from the atmosphere and becomes locked up in carbonates on the surface in a geologically brief time. Lava rock, abundant on Mars, contains a lot of iron in the ferrous oxidation state, a black mineral, and this combines with carbonoic acid to form Siderite, an orange iron carbonate. Mars would have been covered with thick deposits of siderite and other carbonates such as limestone, in a short period, and all of the carbon dioxide would become bound into surface rocks. Thus the condition that a greenhouse is invoked to explain, abundant liquid water, is fatal to a simple carbon dioxide greenhouse in a geologic wink of an eye, a few million years. A geochemical engine of enormous power would be required to force carbon dioxide to stay in the atmosphere rather than combining with the surface, and also to stabilize the atmosphere from collapse in a cold spell. Thus, a host of problems would seem to prevent a simple carbon dioxide greenhouse on Mars from existing long enough for even river channels to form, much less an ocean to exist for a geologically long time.

However, the cosmos is not a simple place; in fact, the cosmos has a tendency to create complications. The problem of a Mars with multitudes of sinuous river channels cut through volcanic rock and an ocean basin on the youngest regions of its surface is a complex one that can be solved by a complex phenomenon known on Earth: biology.

Biology on Earth, powered by sunlight, forms an enormously powerful geochemical engine that pulls carbon dioxide from the atmosphere and water from the ground and combines them to form sugar and produces oxygen as a byproduct. This sugar and oxygen fuels the metabolisms of a multitude of non-plant species who devour plant sugars and combine them with oxygen to breathe back water and carbon dioxide into the atmosphere. The buildup of oxygen in Earth's atmosphere, and the dominance of biology that utilized it, is a fairly recent event in Earth s geologic history, occurring in the last half a billion years. Strong evidence exists that primitive plant life, excreting oxygen, began very early on, Earth, in the first billion years; however, plate tectonics prevented oxygen from building up to significant levels until quite late. As Chris McKay (NASA Ames) has pointed out, plate tectonics on Earth continually exposed lava rock, full of black ferrous, iron, to the Earth's surface. The black ferrous iron would pull oxygen from the air to form red ferric iron, the higher oxidation state, then the moving rock would be subducted back into the Earth. However, as the Earth cooled and became less geologically active, and oxygen-fueled life expanded its foothold, a transition finally occurred where oxygen production overwhelmed its absorption, and the atmosphere of Earth became oxygen rich. Mars had plate tectonics to a degree much less than, Earth. If life began early on Mars, as it did on, Earth, then oxygen levels could have become significant very early in its history.

The effect of a biologically produced oxygen component in the early Mars atmosphere would have been dramatic; for one thing it would have allowed a stable greenhouse. Oxygen, when free, is one of the most reactive gases in the periodic table. When siderite, the ferrous iron carbonate that most carbon dioxide would become bound in a naive greenhouse, is exposed to oxygen in the presence of liquid water, the iron releases the carbon dioxide back into the atmosphere and combines with the oxygen to form hematite, a common mineral on Mars at present. Oxygen has other effects as well.

Carbonic acid is not a strong acid, not even as strong as vinegar, but it reacts with volcanic rock to form carbonates. Limestone is a common carbonate rock on Earth, and is made from calcium carbonate. It is stable, but not to common acids such as are produced by air pollution, like acid rain. For this reason, many classical marble sculptures, marble being a form of limestone, must be protected from acid rain or they will dissolve. The acid rain on Earth is formed by oxygen reacting with sulfur and phosphorus to form sulfuric and phosphoric acids. These acids, even in a weak water solution, react with carbonates to form carbon dioxide and more stable sulfates and phosphates. Thus the oxygen, because of direct reaction with surface rocks, and creation of secondary acids, forces carbon dioxide from surface carbonates to return to the atmosphere. What makes the oxygen to do this is life.

On Venus, the rain is also acid, and ensures that the carbon dioxide stays in the atmosphere. Venus has had a strong greenhouse, apparently, for most of its history. In this, sense, Mars and Venus found something in common, a persistent greenhouse regime. The fact that this was accompanied by sulfur-based acids from oxygen acting on volcanic plumes reminds one, whimsically, of the torrid affair Mars and Venus shared in mythology, with Vulcan, the god of volcanoes, as the third point of a love triangle.

Therefore, biology, by producing oxygen, would have allowed a chemical stable carbon dioxide greenhouse on Mars for eons. Instead of large deposits of carbonates, the surface of Mars would contain sulfates and phosphates, and large amounts of ferric iron compounds such as hematite and maghematite, and because of this, would be bright red. This is exactly what has been found. It has even been found in the most ancient parts of Mars, the southern highlands, where, like the ALH84001 siderites, carbonate beds have finally been found. This confirms that on early Mars, liquid water, carbon dioxide, and low oxygen levels combined to form carbonates rather than sulfates.

An oxygen rich Mars atmosphere would have had other important effects. Mars would have had both a layer of molecular oxygen and an ozone layer at the top of its atmosphere. Molecular oxygen is a powerful absorber of short wavelength ultraviolet, precisely the ultraviolet that destroys water molecules and causes planets to lose their moisture to space. Such an oxygen layer would also preserve methane, a powerful greenhouse gas. Combined with geothermal heat, methane with a protective oxygen layer, and carbon dioxide would have created a powerful greenhouse even in the days of the early dim Sun.

Ozone, produced by oxygen, is an absorber of longer wavelength ultraviolet that is harmful to life outside of the ocean. A Mars with an oxygen-rich atmosphere like Earth would preserve its oceans of water and would have allowed life to leave its ocean, as it did on Earth, and to thrive on the surface. A Mars with a stable ocean and greenhouse, and an oxygen atmosphere, would have had a stable climate.

Ocean coast property in most regions is desirable because it has a more moderate climate. Water has an enormous heat capacity compared to other simple chemicals and this means that oceans tend to stay warmer in the winter and cooler in the summer than the land near them. Combined with the capacity of water to dissolve carbon dioxide and other gases, the ocean of Mars, secure under its oxygen atmosphere, would act as an enormous thermal and gas reservoir to stabilize the climate of Mars.

The problem of the dry ice instability on Mars, the possibility of catastrophic changes in climate and air pressure on Mars, is virtually eliminated by a large ocean as may have existed on Mars. The temperature above the ocean would have remained above freezing as long as the ocean surface remained liquid. Any sudden drop in atmospheric, pressure would have caused the ocean to fizzle and bubble and release a billion tons of carbon dioxide and other gases into the atmosphere to shore up the pressure.

What we would have had on Mars, given just one assumption: oxygen producing life, was a Martian Gaia. It would have been a Mars rich in life, where life itself promoted and stabilized the very environmental conditions it needed to thrive.

Did the old Mars have scenes like this?

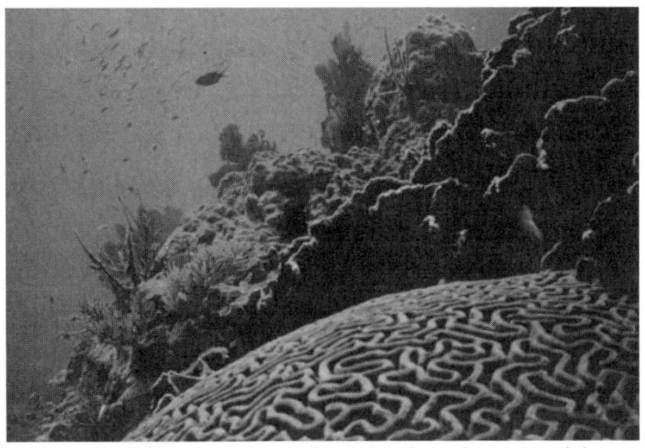

It is impossibel to conceive of a Martian Ocean without life.

To say that Mars may have had a biologically stabilized greenhouse is an understatement. It is possible that Mars did not have a greenhouse but instead a Crystal Palace, a vast planet teaming with diverse life, with an oxygen atmosphere, an ozone layer, a warm ocean. It was robust, and outshone its plodding cousin closer to the Sun, Earth, with its oxygen poor atmosphere. It is in fact possible, as has been suggested by Hoyle, that early Earth was too much of a swampy hothouse to support anything but the most primitive extremophiles. The vastness and diversity that the Martian biosphere could have attained in the billions of years that it existed can only be guessed at. Not only a vast biosphere may have occurred on Mars, but because of the persistence of this greenhouse, as witnessed by the ocean bed on the youngest parts of Mars, what also may have occurred on Mars was evolution.

On Earth, life was seemingly stuck in a holding pattern of one-celled organisms for most of its history. It was only 500 million years ago that suddenly everything living switched to multicellular organisms and predator and prey relationships evolved and an oxygen atmosphere and the metabolism to use it was perfected. On Mars life had billions of years of oxygen metabolism. This was not just enough time for life to gain a firm foothold on Mars, it was time for it to evolve. Everything was set up for Mars and its life, safe in its glorious Crystal Palace, to rule the Solar system, except for one thing.

Chapter 8. The Chicxulub of Mars

"The Cosmos plays hardball."
Pamela Monroe, Asteroid 20-2012 Sepulveda

The cosmos gives life, but it also deals out death. During the age of the dinosaurs a catastrophe of biblical proportions occurred. An enormous explosion ripped the Earth, sending a killing shockwave around the planet. White hot fragments filled the skies, triggering global firestorms. Every green thing was burned. After this, a killing cold and darkness descended over the devastated planet. Almost every living thing perished on the Earth, and only the small and burrowing, the slow and the patient, survived.

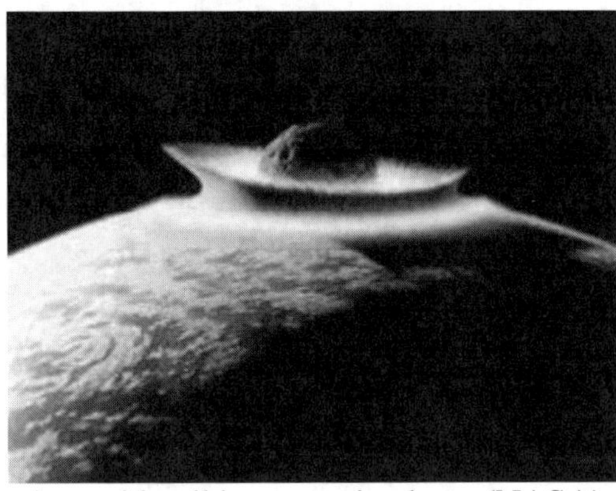

Asteroid striking oceanic planet (NASA)

The great catastrophe that was Chicxulub ended the reign of the dinosaurs on Earth. Its released energy was equal to a million hydrogen bombs going off at once. It left an ash layer that circled the globe and a crater 180 km in diameter, centered near Merida' in Yucatan. It also left in its ash layer a telltale trace of iridium,that told the human investigators millions of years later from where the hammer blow had fallen.

The disaster had come from space. An asteroid, perhaps ten kilometers in diameter, had collided with the Earth. Such collisions are normal for the cosmos. An examination of the Moon through a telescope shows us that over geologic time every planetary surface has suffered overwhelming violence of cosmic collisions.You do not want to be nearby when this normal event occurs. It is like the minor tantrums of nature, tornados, hurricanes, earthquakes. They are also part of the natural flow of events, but that is cold comfort when they strike home or family.

In this cosmos, there are few givens. Asteroid impacts may be the great destroyer of life in the cosmos, and Earth is due, by some accounts, for another Chicxulub event. Thankfully, humanity, for the first time in its history, has the means to both detect and defend itself against another Chicxulub. The scenario of the detection of a large asteroid on a collision course with Earth, and the crisis this will create, has been explored in science fiction, most recently by the author, under the pen name Victor Norgarde, in the novel, *Asteroid 20-2012 Sepulveda*. The school of Mars would teach us to be on our guard continually for such an event and if we receive warning of it, it may be only the lance of Mars that we can wield against it.

Mars orbits next to the asteroid belt, the source of almost all meteorites that strike Earth. Because of this proximity to the asteroid belt, it is reasonable to assume Mars probably receives many times the number of meteorite impacts as does Earth. This is reinforced by the young geologic age of many Mars meteorites, indicating that large areas in the north of Mars are young. Indeed, when the Mars Rovers turned their cameras to the skies of Mars at night and watched the Earth as a bright blue star, they saw numerous meteor trails. The Martian sky at night apparently provides a fireworks show every night of the year. So it was not surprising that one Mars Rover, after exploring only a few hundred square yards around its landing site, stumbled over a nickel iron meteorite lying in the sand. When I asked a meteorite expert at a conference, after a discussion of this meteorite finding, to estimate how much the Mars meteorite bombardment rate must exceed that on the Moon, he only responded “a lot.”

An iron meteorite found near the Opportunity landing site (JPL/NASA)

A possible meteor trail seen from the surface of Mars (JPL/NASA)

Thus, it was perhaps inevitable that Mars had its own Chicxulub event.

The great ages of Mars have now been named and codified. They take their names from the names of the characteristic terrains that formed in the geologic epochs. First is the Noachian Age, the early period of Mars after its surface formed. The Noachian is preserved in the ancient southern highlands. It was obviously a time of warmth and wetness, great bombardments from the sky and great floods of lava. This was approximately the first half billion years, assuming a 4x lunar cratering rate.

Meteor Crater Arizona and yes, it can happen here

The Hesperian Age follows, and makes up the middle age of Mars. Much water flowed on Mars in this time, most of it in the north. The rate of meteorite bombardment eased into a steady lower level, comparable to, but probably 4 times the rate seen on the Earth's moon. This epoch probably lasted 3.5 billion years, assuming a 4x lunar cratering rate. The great Tharsis volcanoes seemed to have formed then as well as the great wrenching apart of the Mars surface that formed the Vallis Marineris.

The Amazonian is the last great age of Mars, and it takes its name from the smooth lightly cratered plains in the north. In the early part of the Amazonian age the climate of Mars became cold and dry. Thus it is in the Early Amazonian when Mars greenhouse may have been irreparably shattered.

On the shoreline of the Malacandrian Ocean, to the east of Cydonia, lies the vast Lyot impact basin. Lyot (pronounced Leo), named after the French astronomer, is 200 kilometers in diameter, thus slightly larger than Chicxulub on Earth. Lyot is the largest young crater on Mars, formed somewhere between one to a half a billion years ago, forming sometime in the Early Amazonian epoch on Mars.

Examination of the ages and characteristics of terrains on Mars demonstrates that Mars suffered strong climate changes during the early Amazonian. Liquid water apparently quit moving on Mars and was replaced by ice. Careful examination of the deep impact crater by James Head at Brown University has also shown that the Lyot crater was not altered by water, to nearly the same degree, as were other smaller craters in the region. This would indicate that Mars experienced a deep freeze that immobilized its water, about the time Lyot formed.

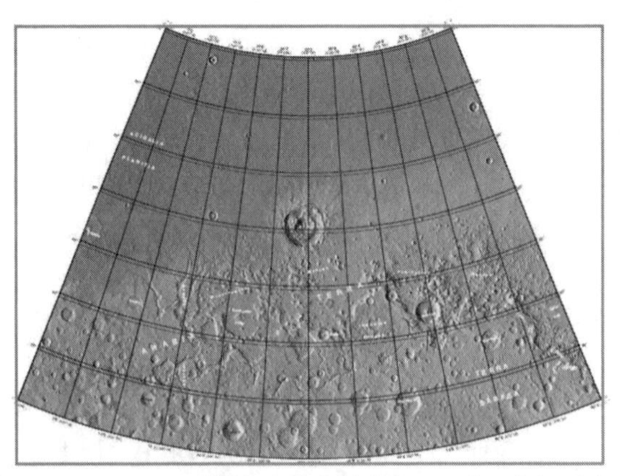

Lyot impact crater at 50N and 330 W, north of Arabia Terra and East of Mare Acidalium (USGS)

Based on its nearly equal size to the Chicxulub basin on Earth, it can be said that the Lyot impact crater would have devastated Mars and plunged it into a deep winter. The ocean would have frozen, cutting off its ability to stabilize the planet's climate. Deprived of that staff upon which Mars greenhouse leaned in times of trouble, the catastrophe foreseen by Sagan and his colleagues would have blossomed fully. Under clouds of dust and soot, the temperature at the North Pole would have dropped below the sublimation point of carbon dioxide and a huge polar cyclonic storm would have developed, sucking in all the atmosphere of Mars. What was left of the biosphere after the explosion and firestorm would have perished from cold and finally asphyxiation. Only the most primitive of living things would have survived in refuges shielded from the direct force of the catastrophe.

Devastated forest in Siberia following Tunguska event

Because of the mixed nature of the Mars greenhouse, primarily of carbon dioxide and water, a strange and terrible thing would have happened following the flash freezing event on Mars: Rather than recovering and thawing, Mars may have swallowed its ocean and atmosphere. Under the frozen and nearly airless sky of Mars after its fall, the frozen ocean would behave more like dry ice than anything seen on Earth. Instead of melting under the noonday sun, it would have turned to vapor, then, being carried by the thin winds, it would have frozen out as water frost on the pile of dry ice on the northern polar cap. This vast migration of the water on the Mars surface to its coldest point would have continued for millions of years forming an ice mound on the northern pole. Under the pressure of a kilometers thick water ice sheet, the carbon dioxide would have been forced into solution in the liquid layer formed on the underside of the ice sheet. Water expands as it freezes, so when ice is placed under high pressures, it seeks pressure relief by melting. This is the same effect that causes glaciers and ice skates to move on Earth. The dry ice forced into the water would form carbonated water. The carbonated water would itself be forced into the soil and rock beneath the polar cap. The carbonic acid would have combined with the deep lava, which is primarily iron and magnesium silicates, to form iron and magnesium carbonates, freeing vast amounts of silica to form sand. Thus the final resting place of the original dense atmosphere of Mars is probably in circumpolar deposits. The freed silica would form the vast sand dune sea surrounding the north pole of Mars. After a million years, the ocean and atmosphere would have been subducted, swallowed whole by the Mars surface. So Mars was irreparably changed by its disaster, and could not recover.

Aiding this strange event was the fact that this disaster occurred late in Mars geologic time and, being much smaller than Earth, Mars had lost much of its reservoir of geothermal heat. The greenhouse of Mars would have retarded the loss of Mars's heat to outer space, but now that greenhouse was gone. The core of Mars would have cooled and frozen, and the water normally kept on the surface by the high temperatures in the deep rock, would now sink deep. So the ocean that once graced the surface of Mars was frozen near the surface, or liquid kilometers down. There, this water may have awakened one last bizarre phenomenon.

Lyot impact crater on Mars is 200 km across, larger than Chicxulub (NASA)

Impacts on Mars had probably occurred before with less violence, but even if the world had partially frozen, Mars generated enough geologic heat early in its life to help thaw itself and force water to stay on the surface. The ocean would have recovered, and become free of ice, released gases, and Mars would have revived itself. However, late in its history, the loss of its greenhouse could not be remedied, so Mars slept. However, the original impact that created Lyot also threw material across the solar system.

Every dark cloud has a silver lining. Just as the later Chicxulub event was the doom of the dinosaurs but a historic break for the mammals, so the Lyot impact on Mars may have been the occasion for strange good on Earth. The timing of the Lyot impact appears to coincide with the Precambrian explosion on Earth. It is possible that the Precambrian explosion was in fact an invasion from Mars. This invasion took the form of a shower of Mars meteorites bearing rugged spores of multicellular Martian life, still primitive by modern Earth standards, but revolutionary on Earth.

So Mars may have lived, and its biosphere become glorious, in its Crystal Palace by the sea. But then the killing star came, and swept it all away with first fire and then ice. In its frozen state Mars's ocean was taken from it and its nakedness shed its last reservoirs of life-giving warmth. Splendid was the life of Mars, and terrible was its death.

Northern Mars showing location of Lyot Crater (NASA)

Chapter 9. The New Mars Synthesis

"There is nothing so extraordinary as the commonplace."
Sherlock Holmes, A Study in Scarlet

It possible we now know the core story of Mars, a story of beginnings and endings, of both life and death. The riddle of Mars's appearance, its many ancient water channels in what is now a frozen desert, its vast canyons exposing many layers going to kilometers' depth, the redness of the layers and its surface, the dry basin of an ocean, was like a story inscribed on stone tablets in some strange forgotten language. We can now translate the inscriptions on the stones of Mars in large part. The key to the translation is the word "life." The translation of the text follows from this key word.

Life is the one entity that allows us to use Ockham's Razor at Mars. With this one entity, the whole puzzle of Mars falls into place.

Water frost on Mars Utopia Planitia (JPL/NASA)

Mars, like Earth, was apparently the dwelling place of life early in its history. The source of this life was most likely spores that have traveled between stars since some forgotten time in cosmic history. So Mars and Earth were both seeded from a common source. This concept is not new; it is called Panspermia, from the Swedish scientist Arrhenius. The idea even predates him, going back in some form to Greek philosophers. When combined with a steady state universe by Sir Fred Hoyle, it gives a picture of a universe that continually creates itself with life imbedded, a concept of a whole cosmos with life as integral part, existing forever in the past and future. So life predated the planets of this solar system and perhaps the galaxy itself. This would mean that Mars has told us the same story our own planet has, that we dwell in a living cosmos.

Life on Mars grew as Mars cooled earlier than Earth. It was smaller, so had less geologic heat. This was the Noachian epoch. The hammering of the leftovers of its accretion on its surface was less intense. It was farther from the early weak light of the Sun, so its initial dense greenhouse produced by carbon dioxide produced only moderate temperatures compared to the hothouse of the early Earth. Mars's core was hot and generated a strong magnetic field to shield it from cosmic rays and solar flares. The scattered waters of Mars were gathered into one place in the north, abandoning, to a large extent, the highlands of the south. A hydro-cycle of evaporation and rainfall began which filled rivers that emptied into the northern ocean. As its atmosphere was shrinking due to formation of carbonates on the ground and loss to space due to ultraviolet from the early Sun, a transition was made where temperatures dropped to be very favorable to complex micro-cellular life with nuclei and a diverse set of enzymes for surviving in a wide variety of habits now forming on Mars. Thus did the Noachian age end.

This advanced microbial life was very fertile and vigorous and spread through the northern ocean and rivers that fed into it. In keeping with Mars's smaller geologic heat engine, tectonic movements of stony plates ground to a halt after a short period. This meant oxygen could begin building up its atmosphere from photosynthetic cyano-bacteria, since fresh lava rock was not being renewed on its surface to be a sink for oxygen. Mars began to thrive. This was the Early Hesperian.

As oxygen built up in the atmosphere, it began to determine the type of life that would dominate the Mars biosphere. This life was vigorous, powerful, aggressive, and always hungry. The oxygen created a protective double layer of UV protection, a molecular layer to stop the hemorrhaging of water, an ozone layer to protect life that crept out of the water onto land looking for food. The oxygen made Mars a slightly sour place, where carbon dioxide could find no resting place on the surface, and was locked into the sky. Mars's atmosphere stabilized in pressure. It was secure from the Sun's UV, and it was secure from chemical combination with the soil. The magnetic field of Mars began to die as the core of Mars grew colder; however, the oxygen layer created a dense plasma that trapped magnetic flux and thus shielded the atmosphere from the erosive bombardment of the solar wind, even as the core magnetism faded. Mars had suddenly become a good, stable place to live.

Life on Mars was running the place now, like the planet was a giant organism. It was the young days of the Martian Gaia. Photosynthetic bacteria made oxygen to sustain the atmosphere. Other bacteria ate the sulfur and phosphorus from volcanoes and burned it with the oxygen to make energy and sulfuric and phosphoric acids to keep Mars pickled, and thus sustain its carbon dioxide greenhouse. Other bacteria fermented the sugars made from photosynthesis and made methane to buttress the already formidable carbon dioxide and water vapor greenhouse. The oxygen and ozone layers protected the methane from the ultraviolet that would destroy it quickly, so instead the methane oxidized slowly. Other bacteria opportunistically ate the methane-making bacteria when they got too numerous, while other bacteria ate the methane itself when it became common, and turned it back into carbon dioxide and water from whence it came. They used the energy from this to live and move and at night would emit light like fireflies, to proclaim their complete reversal of photosynthesis. Some bacteria began to eat the plant bacteria, and later, they ate any other bacteria that they encountered. Thus began an arms race on Mars. This was the middle Hesperian. Things evolved quickly, and plants invaded the land.

Does fossil Mars life look like this?

Conditions on Mars,before the fall?

The biosphere on Mars began to diversify and evolve. Bacteria quit merely congregating in mats where there was food, banded together in small platoons to hunt for food together. In unity there was strength, and movement, and successful hunting. Other bacteria banded together for protection from these roving units, and welcomed plant bacteria into their protective spheres to feed themselves in return for protection of the plants. Some units of bacteria began to get really organized and to develop specialist cells that did nothing but propel the colony by manning oar-like flagellum, others began acting as sensors of things to eat. Some became grappling hooks to grab and pull bacteria and other smaller units into the killing zone of the hunting units. Some began to manufacture venom to kill whatever other units were encountered so they could be eaten more easily. Soon the ocean of Mars was an armed camp of prey and predator organisms, competing for food and for good places to live. At the water's edge organisms crept out onto the shore to outflank their competition or escape the mouths of those who could not follow. They found the shore more mellow and accommodating and they journeyed inland. They journeyed as far and wide and as high as they could. The great Martian biological enterprise was rolling and nothing, almost nothing, could stop it from attaining a highly advanced state.

How far did life advance on Mars in 4.0 billion years?

Eons past, the troublesome asteroid belt next to Mars would occasionally visit the planet with terrible cataclysms, and life would be forced to retreat back into its burrows and to the oceans to regroup. But the ocean and the greenhouse held. Carbon dioxide might condense on the poles sometimes, but reinforced by the warm ocean and the methane as its backup greenhouse gas, Mars's climate would recover. Life would thrive again, this time even more tenacious and resilient.

Mars favored life that could survive the occasional cold spell. Life that could construct its own Noah's ark of a spore like covering, impervious to cold, to low pressure, to radiation that would leak through the occasionally thinned atmosphere, came to dominate the place and when this life found itself suddenly grounded on a Mount Ararat after the flood, it would aggressively sally forth and begin hunting for food and places to live. Life was favored that knew to be afraid and retreat to a cave when bright flashes occurred and winds blew, and perhaps take a nap. Successful life was the first to emerge and devour the frozen dead and have a new crop of offspring to dominate the thaw. Mars was where the survival of the fittest meant the cultivation of good habits, and a keen awareness of when to run for cover and go dormant.

Things might periodically go into a deep freeze on Mars.

Observers despaired of any such development on the nearby hothouse of Earth. Earth was stuck in a deep rut. Life of Earth consisted of gentle colonies of bacteria in the ocean, without malice or ambition. The living forms on Earth were generally afraid to play with oxygen, viewing only as a dangerous waste product of photosynthesis which fortunately was so reactive that it combined with the very rocks before it could reach dangerous levels. The rocks continually renewed themselves, like a vast tectonic conveyor belt. The great struggle of life consisted in creeping fast enough to avoid tectonic subduction. The Sun was now running strong, and the atmosphere clear enough that photosynthetic bacteria on Earth could supply a steady stream of compost for the gentle anaerobes in the dark watery depths below to digest. The Earth's dry surface was bare rock for the most part, seared by ultraviolet energy to discourage any trips to the shore. Occasionally asteroids would escape the belt and get past Mars and tear up the place. The inhabitants of the oceans barely noticed.

However, year after year, the levels of oxygen rose on Earth, life's oxygen engine now outstripping the geologic engine in its ability to consume it. With no aggressive eaters of plants, plants took over the near ocean surface, driving the anaerobes to the depths of the sea by the rising oxygen levels. Oxygen was the ultimate poison that plants could make to protect themselves from any feeble eaters of the sugar they made. This explosion of plant life began to be a serious sink for carbon dioxide on Earth. Carbon dioxide was disappearing into the sweet waters of the ocean in the form of sugar and more slowly into the formation of ferrous iron carbonate. This meant the temperatures on Earth were dropping to the level regulated by the thermodynamics of a water hydro cycle, where more advanced bacteria could prosper, and the simple extremophiles were more and more exiled to hot springs and the deep underground where they did not have to compete with their more sophisticated cousins.

So the worlds turned in their courses around the Sun. Earth slowly cooled and its atmosphere became more oxygen rich, and the level of CO_2 dropped as it was captured into sugar. On Mars the biosphere pulsed with activity and continued to rule the planet. Then suddenly something happened to change everything.

The asteroid belt, as if jealous of its favored neighbor Mars hurled at that planet an asteroid much larger than anything in the last four eons. Living next to the asteroid belt, Mars's luck had finally run out. The impact was near the northern ocean shore, concentrating its force and dust cloud in the north. The ocean's surface froze, cutting off its help from the atmosphere, and the whole atmosphere condensed into an awful polar vortex in the north and finally in the south. Mars's crystal palace, which had stood resiliently through so many previous small impacts, was now shattered irreparably. It was too much for even the rugged Martian life to endure. All over Mars, life went rapidly into a hibernation from which it would never awake. But in the very violence of the Lyot impact, and its cruelty in hitting the center of the biosphere on Mars, the north, there was a bizarre hope. For many primitive Mars organisms, secure and asleep in their little Noah's arks, a new dawn would come, on Earth.

Mars global firestorm, the aftermath of the Lyot impact

The arrival of a vast shower of rock from the center of biology on Mars was a cataclysm on Earth. The sky was soon filled with shooting stars and the ground with impacts. Earth's Moon absorbed its own share of Martian rock, which remained where it lay to tell the tale of the event. On Earth, it looked like an invasion from outer space and it was. Out of the rocks, as soon as they cooled in the ocean, sprang a multitude of rugged, voracious and sophisticated Martian organisms. They were springing to life on contact with liquid water and warm temperatures as their ancestors had done for eons on Mars. Fortunately, for the refugees from Mars, the still heavy carbon dioxide atmosphere on Earth was close enough to what they were accustomed, and the oxygen level was just high enough for them to breath. It was a rugged transition, but they were rugged. Mars had kept a rough school, and only the hardy survived there, for every million Martian organisms that died on emerging into the alien environment, a few dozen were tough enough to survive the new environment of lower ph and lower oxygen. For the

Martian organisms that survived, Earth was a sweet place, and a happy hunting ground.

The invasion from Mars took Earth by storm, just as the invading Martians in H. G. Wells's classic found the humans nearly defenseless and disorganized, so the Martian organisms found the Earth occupied by sweet helpless plants, and single-celled predators who were averse to oxygen. They conquered the place, like rabbits conquered Australia. The plants of Mars, also, with their adaptation to living on land, awoke vigorously and took root. Soon oxygen production was soaring, driving the anaerobes to the deep ocean bottom mud. Carbon dioxide levels fell, cooling Earth to a more Martian level, as if the Martians had turned the tropical paradise of Earth into one giant air-conditioned hotel.

Here and there Martian life forms and Earth life, being long distant cousins, formed symbiotic communities. The terrestrial cells knew the best way to live, and the Martians needed to eat. Native Earth predators, a small tribe who were not afraid of oxygen, found chinks in the Martian armor and soon began looking for the newcomers as prey. In the ocean, every bacteria colony that could form a calcium carbonate shell to protect itself from the marauding Martians did so. Those who were not lucky enough to have this mutation were not heard from anymore. Earth's ocean became a region of castles and fortresses against the invaders. Carbon dioxide was suddenly vacuumed into the oceans for the construction of fortifications. Earth's greenhouse effect fell to its lowest point in history. Where before Earth had rewarded the stable and plodding life forms, now Earth gave emergency land grants to those who were opportunistic and prone to daring experiments in the face of biologic and climatic disaster. Earth's biosphere staged a counterattack moving its own oxygen breathing predators to the front lines of the ocean shore, and armored bacteria colonies appeared also on the coasts. The Martians, now numerous, and themselves relentless opportunists, adapted also. After a wink of geologic time Earth had been transformed biologically.

The biota of Mars had landed and conquered, then been pushed back, and then become part of the landscape. Where the precise genetic lineage of Mars was perhaps difficult to follow, the spirit of Mars life was now part of Earth's heritage. Mars of the old was now dead, a frozen husk of its former self, where only in a few refuges did life cling to existence. But Mars biology had transplanted its soul to Earth, if not its actual substance. It was a voracious and opportunistic, high-rate-of-metabolism, soul. It was the soul of the predator.

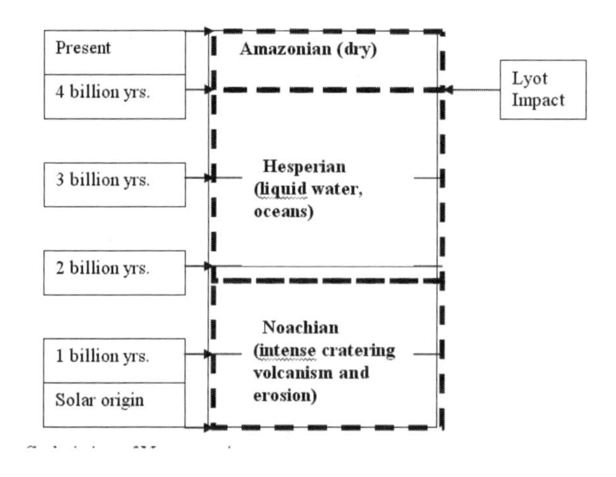

The Geologic Ages of Mars, assuming an approximately 4x lunar cratering rate

Chapter 10. The Twilight of Mars

"In Surtur's grasp the Sword of Revenge blazes, adding a blood red color to the twilight of the whole world."

Old Norse saga

In Gotterdamerung, "the twilight of the gods," or Ragnorak, the "doom of the gods," the old Norse legends, the end of the world comes after the winter of winters when the whole world froze. The forces of evil and the wicked dead gathered on the plain of Vígríðr and were met in battle by the gods of Asgard leading the army of the valiant dead from Valhalla. Tyr falls in battle after slaying his great foe, the hell-hound Garm. Likewise, Thor battles a great serpent to the death. Then came Surtur, from Muspelheim, the land of flame, and in a rage wields his flaming sword and burned the whole nine worlds with fire. The Earth is turned to ashes and sinks into the sea. Only in the forest of Hodmimir were living things preserved. So the world first freezes and then is seared by fire in one final act of rage.

Ragnarok: The Last battle of the Gods where Tyr fights the great hell-dog Garm

The end of Mars was not simple. It froze and its atmosphere rained out onto its poles, but the dynamics of catastrophe did not finish there. After Mars perished in ice, a final mysterious disaster of fire apparently occurred involving phenomena known only recently on Earth. The evidence of it had been known for years, since the Viking landings, but the questions the evidence provoked were unspeakable, so it remained unspoken. Not until a discovery in Africa had been fully understood could a hypothesis be formed and presented.

In Africa, at a place called Okla, are deposits of almost pure uranium dioxide, also called yellow cake. The deposits from 1 billion years ago, before even the Precambrian explosion, are caused by the interaction of bacteria and ground water. As this yellow cake was being mined it was discovered that in at least eleven different sites, each the size of a dining room, nature had begun its own exploration of nuclear power.

Natural fission reactors, moderated with light water like the usual power reactor in the United States, had run in cycles of hours for close to a million years. The reactors had run because the uranium was nearly pure, was soaked in ground water to slow down the neutrons and make them more reactive and also because 1 billion years ago all natural uranium was 3% U-235, whereas now it is only 1% due to radioactive decay. The natural reactors had bred plutonium from the U-238 that was mixed in with the U-235, and to the amazement of the scientists who studied it, had not exploded but had run in cycles of activity and dormancy. They had done this because their small size allowed water to diffuse in, cause the reactor to go critical, become very hot, turn the water to steam so it would escape, and then without water to act as a moderator, would shut down and cool off, so the cycle could be repeated again. Like any other nuclear reactor it had generated high level radioactive waste, but had naturally sealed it in aluminum phosphate clay so it never escaped. That this happened not once, but at many locations in the uranium deposit suggests that this is a process favored by nature, and has probably occurred at many other places on Earth in the past. However, it apparently also occurred on Mars, but far more catastrophically.

Thor dies while slaying the great serpent at Ragnarok

That a violent nuclear event occurred on Mars in the distant past is dramatically evident from the pattern of isotopes in the Martian atmosphere and the pattern of radioactive thorium and potassium on its surface.

The Mars meteorites contain little uranium or thorium, compared to Earth rocks. Uranium is named for the Roman sky god Uranus, and Thorium for the Norse god Thor. This presented a puzzle, because the Russian Mars probes had found Earth-normal abundances of both on Mars. This Russian result was confirmed by the Mars Odyssey. This meant that the Mars uranium and thorium had to lie in a thin layer on the surface, rather than in the rocks below it. Radioactive potassium was also hyper-abundant on the Mars surface, in a pattern that matched that of the thorium. It was as if some process had spread dust rich in thorium and uranium all over the planet, and had also irradiated potassium with neutrons in the same pattern. Then the maps arrived that showed the thorium and potassium had spread out form one area of Mars, complete with another hot spot at the antipode, the place where any shockwave would refocus as it moved around the globe. The area was called Mare Acidalium. It forms a large, dark, burned looking spot on Mars.

Mars Odyssey with a boom-mounted Gamma Ray Spectrometer (JPL/NASA)

The atmosphere of Mars held strange clues. The pattern of xenon abundance in the Martian atmosphere is different from any other planet or meteorite. Xenon, a heavy element, has many stable isotopes, and the pattern of their abundance is a fingerprint for the processes that have affected the bodies where it occurs. The xenon 129 appears to come from a large violent reaction like a nuclear bomb of tremendous size. Another rare gas on Mars also has disturbed isotopes. The krypton 84 found trapped in Mars meteorites appears to come from the neutron irradiation of a large area of the Mars surface. The argon 40 abundance was also unusually abundant and must have occurred from the decay of a large amount of radioactive potassium.

The pattern of thorium and radioactive potassium is shown in the plume from Mare Acidalium, spreading out in every direction. Another hot spot at the approximate antipode suggests a violent explosion in Acidalium that spread debris and shock waves clear around the planet. The author is grateful to Dr. Edward McCollough, for pointing out this antipodal hot spot and its significance.

The trigger for this bizarre and disastrous event may have been a shift in groundwater that occurred on Mars after the collapse of the Mars greenhouse. The combination of a buried asteroid from the early period of the planet that was rich in uranium and thorium, and a dramatic invasion of ground water deep into the asteroid's resting place when climatic collapse occurred created conditions leading to a second planetary disaster. This concept was well received at the conferences where the author presented it. One researcher remarked, "if natural nuclear reactors occurred on Earth, they must have occurred on Mars." True enough.

Uranium and thorium are two elements that tend to occur together in nature and both are capable of nuclear fission. Thorium is more abundant, by a factor of three. The invasion of ground water into the deep deposit of uranium may have made the reactor go critical, but because thorium, upon neutron radiation, becomes U-233 the reactor bred more fuel the longer it ran. U-233, like plutonium, the favorite element for nuclear bomb builders, has the nasty property that it becomes more reactive as the energy spectrum of the neutrons becomes higher. This is the opposite of U-235, which loses its reactivity when the neutrons lose moderation. This means that if a reactor got going, bred a lot of U-233 on the thorium, and then lost its water moderator due to steam, it would not shut down like the reactors in Africa, it would instead run away and become a Chernobyl. If this occurred in a deep deposit, the reactor would be held together by the kilometer thick overburden much longer than if it was near the surface. This means it would generate far more energy in the seconds before it exploded. This means that instead of a Chernobyl, it would be a doomsday bomb.

The invasion of ground water could have been triggered by the ice mound that formed after the Lyot impact that cooled the planet and forced its water underground. Thus the collapse of the Mars greenhouse, and the rapid cooling of the planet, causing the ocean to sink, triggered the final catastrophe that scorched the planet and may have destroyed much of the life that may have survived the greenhouse collapse.

If there is any problem with this scenario, it is that the Acidalium region shows a huge scorch mark, but not a crater. It was for this reason that, when the author wrote up the hypothesis for publication in a scientific journal, it was quickly rejected. The article is included here as an appendix. The solution to this problem will require careful study and sample return of the region, as it possible that the region was so hot that it melted the rock and filled in the crater quickly, or else it filled by dust. A depression exists in the center of the scorched area, but it is shallow. So we have a fine, simple story, but no crater, only some disturbed looking ground.

Perhaps something else happened. Tyr, the Mars of the Norse pantheon, was not only the god of war but also of the Allthing, where everything could be discussed. This thing was before unspeakable, but now we must speak of it.

Perhaps the explosion occurred above the surface, not buried beneath it. This would account for the absence of the crater and the pattern of xenon isotopes, which matches a violent explosion and not a nuclear reactor. That scenario is not called a natural nuclear reactor; it is called, in nuclear weapons terminology, an "Airburst." It almost appears as if someone was not content with Mars dying a death of ice, but decided to sear the planet with radiation as well, and therefore dropped a huge bomb on it. Cydonia lies close between both Lyot and Mare Acidalium. However, this is a much more complex hypothesis than natural nuclear reactors. It involves not just great intelligence, but also great malice. Therefore, the absence of a crater means that we must leave open the possibility that Mars was the site of interstellar genocide on a planetary scale, like the planet Alderan of Star Wars. The idea that the whole of Mars may have been the site of mass murder is unimaginable except for the fact that we have seen such things on Earth. We should investigate this matter quickly. However, we also need to gather more data while considering all possibilities, and not dwell on negatives until we have to.

Surtur wields his flaming sword

Valhalla Burns

Therefore, Mars died two times, once a death of ice and once a death of fire. In both cases the scale of the disasters staggers the imagination. Welcome to the cosmos my friends, enjoy its sights, but remember it is as harsh as it is beautiful. Learn well at the school of Mars: in the cosmos, are the quick and the dead, so be quick.

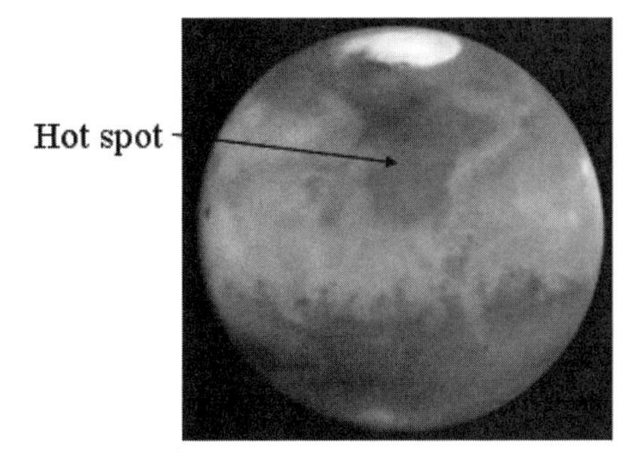

Mare Acidalium, where whatever happened, happened.

Chapter 11. The Endgame of Mars

"You and I both know Stu, that proof is for fools, that is why we stay light years ahead of them."

Cassandra Chen, Morningstar Pass, the Collapse of the UFO Coverup

"God is subtle, but he is not malicious."

Albert Einstein

"We got dead algae from Mars—Don't nobody panic!"

The author, to his colleagues, on the discovery that CI were Martian

Mars probably lives and the bio-terracentric viewpoint of the cosmos is probably dead. Life on Mars has not hidden itself from us; it has in fact gone to great lengths to make its presence known. We just had to know where to look, and we had to be ready to find it.

The methane plumes of Mars were first noted in the early part of the millennium by ground based telescopes and then confirmed by orbiting probes. This is testament to the major effort humanity is exerting in the space around Mars, where numerous probes from various nations now orbit, and the increased sophistication of Earth-based telescopes. The fact that this methane could be seen at all was a testament to the increased sensitivity of instrumentation over that carried by by Mariner 9 in the 1970s, which could have perhaps detected it if had been slightly more sensitive. We have progressed, enourmously, both technologically and conceptually since then. We are much more open to finding life now, we are in different times now.

The Mars methane plumes begin in the northern spring of Mars when the ground begins to warm and peak in their output in late summer. Mars is putting out approximately 19,000 tons of methane per year. That is a lot of methane. Mars appears to be dead geothermally. No areas of the Mars landscape, even its volcanoes, are putting out heat. This means a geothermal origin of the methane is ruled out. The only thing known to make such large amounts of methane other than geothermal mechanisms is biology, specifically anaerobic bacteria consuming dead organic matter and converting it to methane to make energy. This occurs in swamps on Earth all the time and in the intestines of cows. The vast amount of methane on Mars is startling. It is as if life had fired a flare to announce its presence. The plumes come from a water rich area of Mars near the equator called Arabia Terra. The U.S. is already planning to land a new nuclear-powered Mars science rover to the area to investigate, a last move in a cosmic chess game.

In 2010, it is even possible that the Mars Opportunity Rover, approximately 500 km away and in perfect working order, will be dispatched on an epic trek, at a breathtaking 1 km a day, to reach the edge of Arabia Terra and use its instruments, though not optimized to look at methane, to give us knowledge on the ground. If it is dispatched quickly, and survives the trip, and the road is good, it will reach Arabia Terra before a new improved rover can land there in 2011.

The Opportunity has been on Mars for five years, and its team of operators, brilliantly led by Steve Squyres, have become masters at living and moving robotically on the surface of Mars. Even if Opportunity cannot detect methane directly, it can look at the ground where it originates. Its infrared spectrometer may also detect other organic gases. Then the fully optimized rover will land, and if properly equipped, will satisfy the last skeptics in the Mars life debate. The human race is being treated to the final acts of a great age-long scientific drama that has been evolving in real time on Mars and Earth.

Mars swamp gas, methane sources are in Arabia Terra (NASA)

This last act of the life-on-other-worlds drama began in the 1980s with the discovery of Mars meteorites and digestion of the Viking Mars data. Adding to this was the Cydonia data, also driven by Viking data, that sharply increased public interest in Mars, making Mars "cool" again, and worthy of more funding.

The Mars life experiments, particularly the labeled release experiment of Gil Levin (founder and CEO of Biospherics Inc. and member of Viking team), were reexamined, and the various arcane inorganic chemical explanations invoked by the lunar Mars camp were dispatched one by one. The author, noting the dogged ferocity of Gil Levin in this systematic scientific slaughter of the inorganic scenarios, reported a nightmare to him in which, "I was the proponent of the last remaining inorganic mechanism for the Labeled Release results, waiting for Gil Levin to arrive with his siege train." With superhuman effort, Levin forced NASA to change its verdict on the Viking life results, in 1988, from "negative" to "inconclusive."

The proponents of lunar Mars had emphasized two salient points in their argument against life on Mars: the absence of detectable organic matter in the soil of Mars and the dryness of Mars. Life, as we know it, requires liquid water and Mars had none. Those who believed Mars had life, at least in its past, began to focus on these two objections.

The dry Mars objection was the first to fall, at least for Mars past. Examination of the Viking imaging data revealed that flood water, enough to fill an ocean, had moved across the Mars surface in the past. Some argued that the channels were carved by some other liquid, such as methane or liquid carbon dioxide. However, water, composed of hydrogen, the most abundant element in the universe, and oxygen, the third most abundant element, makes water the most abundant chemical compound in the universe. Together with the fact that all plausible models for Mars's past atmosphere were either too hot for liquid methane or too low in pressure for liquid carbon dioxide to form, the abundance of water ruled out all alternatives for forming channels. Viking also revealed that the northern polar cap of Mars was water ice, not dry ice, as had been previously supposed. The ice cap was found to be apparently kilometers thick and held enough water to cover the planet to a depth of several meters. So Mars has plenty of water, both in the past, when it was liquid, and in the present where it was mostly frozen. By the late 1980s ancient Mars had acquired an ocean. This meant life could have begun on Mars as it did on Earth.

Mars Northern Polar Cap (NASA)

The Mars life debate had not only changed with our concepts of old Mars, but also with our concept of life and its origins. Life was viewed as a rather fragile and unlikely phenomenon in the 60s and 70s, but by the 1980s life was viewed increasingly as amazingly rugged, tenacious, and ubiquitous in time and space. The oldest known rocks on Earth, 4.0 billion years old, had been found to contain signs of life. This meant life on Earth was thriving as soon as its ocean existed. Life was also being found where it was thought impossible to exist. This meant life was either very likely to arise from the primordial soup, or as Sir Freddie Hoyle had long insisted, life arrived on Earth as spores from space. Increasingly, the old scenario of life spontaneously arising on Earth from a collection of organic chemicals in an early ocean with a methane and ammonia rich early atmosphere above it appeared unlikely. The early Earth atmosphere was probably more like the present on Venus: carbon dioxide, nitrogen, and water rather than the old Titan-like model of methane and ammonia. Since early Mars and early Earth looked increasingly similar, and life had obviously started early on Earth, by the late 1980s, there seemed no scientific reason to exclude life from early Mars. And, given, the newly understood tenacity of life, if life had begun on Mars, surely some it would survive to the present. So the hunt began in the newly recognized Mars meteorites for signs of Mars life, past or present.

In the age of Mars probes, the search for life on Mars became a search for liquid water, past or present. "Follow the Water" became the new slogan of the Living Mars partisans, who now had turned the objections of the lunar Mars people on their heads. If liquid water is required for life, then we will find it, and life with it. Following the spectacular loss of the Mars Observer, and partly fueled by the Cydonia debate, a renewed resolve to investigate Mars formed in the public and at NASA. From being a dead, has-been world in the late 70s, Mars after Mars Observer was now a center of plots, rumors, struggles, and burning public interest.

The Mars Pathfinder arrived in 1998, inspired in part by the successful Russian Moon rovers, and enthralled the public with views of Mars's surface and close-ups of interesting Mars rocks. Pathfinder, landing on the former ocean floor, found evidence of massive amounts of water and geologic activity in Mars past. Mars was now smoking hot. The Mars global Surveyor also arrived in 1998 in orbit and discovered vast beds of sediments, apparently laid down in Martian lakes and oceans that stretched for hundreds of kilometers. Mars water seeps were even suggested in some areas, as if Mars still had a water table in places and water was seeping to the surface.

The twin peaks of Mars (NASA)

The sediments of Mars (NASA)

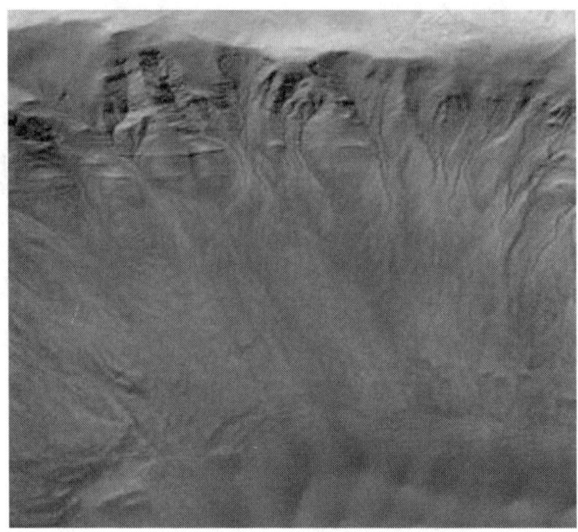
The seeps and gullies of Mars (NASA)

The Mars Odyssey arrived at the next Mars opposition in 2000. Its keen eyes caught a certain sign of water in Mars past. A large area of bulk hematite, an iron ore on Earth, was found. In a persistent and skillful waging of a scientific campaign, a new Mars rover, Opportunity, was fired straight at the hematite area in 2002 and scored a bulls-eye both geographically and scientifically by confirming the from-orbit sighting of hematite. It also confirmed that it was a type of hematite formed in liquid water that had existed in Mars past. In a bizarre and wondrous scene, worthy of a science fiction movie, the rover found that it had landed in an area of hematite spheres, each a few millimeters in diameter. These were termed "blueberries" for their size and color. The spheres covered the ground to the horizon in every direction, as if the region had formerly been covered by a herd of iron eating rabbits. Similar spherical hematite, called "pesolite" (pea-like), is found on Earth in geologic beds from the dinosaur era, and appears to require organic matter or perhaps even bacteria to form. Similar spheres are found formed in Russian bogs, where they are formed by bacteria colonies.

However, for a significant minority in the scientific community, the opinion was already forming that life on Mars, past, present, and future, is a reality. Mars had past life because it was like the Earth in the past, and early Earth was nothing special. Mars had life now because life is tenacious and wants to survive, and it can always find some place to hide when surface conditions become hostile. Life will be on Mars in the future because we are going there.

For many of us, the ALH84001 life results are compelling, and have not been refuted despite the labors of an army of scientists, eager to prove that early Mars was dead.

In the long endgame that this debate has become, Mckay and his team have continued the scientific fight, now ignored largely by the press. McKay's critics, having run out of scientific arguments, have now fallen back to simple disbelief. A NASA official conceded that the debate had run its course largely because, "the bar (of proof of exobiology) is now set so high."

So, set the bar high enough and it becomes a roof, and you can hide under it, but from what?

For the author, the moment of truth about Mars life came when I found that the CI carbonaceous chondrites were most probably from Mars, and saw that despite the best efforts of many scientists to disprove this, it remains a solid result. The CI are full of organic matter and also microfossils. They are 4.5 billion years old, like ALH84001, and contain the same mineral called brunnerite, an iron carbonate, which is found only in CI and ALH84001. Life in the early Solar system was someplace close to Earth, but not of it. That someplace was apparently Mars.

Microfossil algae found in a CI carbonaceous Chondrite (NASA)

The CI also were found to contain small grains of lava, and these also were found to have Martian oxygen isotopes and mineralogy.

As far as the author is concerned, the ALH8001 results largely prove life on ancient Mars and David McKay and his team should get the Noble prize. An article arguing, once again, the CI origin on Mars, and thus confirming the McKay findings, forms an appendix to this book. When these results are added to the life experiment results from Viking and the methane plumes now seen over Mars, the case for life on present Mars is also very strong. However, the realization that something is true, is ultimately a decision, and when humans make decisions about a matter they often access far more data than is on the table in front of them. They must consult whether a decision to believe one thing contradicts many other things that they believe. To decide that humanity is not alone in the universe based on data from Mars is such a decision. You must choose life.

Chapter 12. The Moons of Mars

"The larger the island of knowledge, the longer the shoreline of wonder."

Isaac Newton

"Where the hell are they?"

Enrico Fermi

So now we know the approximate story of life and death on Mars. As with most journeys of scientific exploration, we arrive at the end with more questions than answers. We have gone to the first terrestrial planet in the cosmos we could reach and probably found life -what does this tell us? The laws of physics we know on Earth seem fully operational on Mars, and these laws of physics allow biology on both planets. The same laws of physics seem to be uniform in the rest of the cosmos, based on observations of the radiation, both optical and otherwise, that we observe from the rest of the cosmos. Because we have tried to derive truths concerning the whole cosmos, from what we have found on two planets, Earth and Mars, the implications of our conclusions must, by definition, be cosmological in their enormity.

Mars with its two moons, Fear and Terror (NASA)

Let us assume for this discussion that Mars has life, what now? The first thing our discovery on Mars tells us is that we are not alone. Mars is the first Earth-like planet in the cosmos upon which we could look for life, and we have apparently, by several lines of evidence, found it. This means, since we have now found many planets orbiting other stars, that planets like Mars and even Earth are to be found elsewhere in the cosmos. Our discoveries on Mars mean that life is probably also to be found on them. Based on our findings on Mars we probably live in a universe full of living things on innumerable worlds.

An important concept in science is the Assumption of Mediocrity. This is the idea that the Earth and all it contains are not exotic or unusual in the universe; rather we are commonplace and unremarkable. This is, of course, only an approximation; we are unique and remarkable in our way, no doubt, just as each of us is unique and remarkable in a world full of human beings. Based on a life finding on Mars, the cosmos is probably full of unique and remarkable living things. Every kind of life found on Earth, both known and unknown, is probably to be found elsewhere in the stars. If one finds a thing in one's own yard and then looks in the yard next door and sees it there also, one must conclude it is neither rare nor miraculous: it is commonplace. If we find life on Mars, then life is a common thing in the cosmos. This means the cosmos is, despite occasional catastrophes, generally a hospitable place for life. As further evidence of this conclusion, we are having this conversation.

The second thing our Martian discovery tells us is that we most likely share the cosmos with intelligences like ourselves. This follows from the evidence on our planet of a process of evolution of biology to produce organisms of higher and higher complexity as geologic time has progressed. This evolutionary process is seen even in human affairs, and is called progress. Progress has occurred that is technological as well as cultural. The average human has a better, longer life than they would have centuries ago. We can assume similar life, intelligence, and cultural progress has occurred on other planets that orbit other stars. This means our future, if we are quick and diligent, will resemble the universe of Star Trek. We will have neighbors and we will have exchanges of ideas and goods with them. We will progress together. Humanity, if it is quick and diligent, will take its place in a community of peoples of the cosmos. However, this second thing we have learned, that other intelligences like ourselves are out there, has its own special baggage. This "baggage" has contributed to the crisis of science we have experienced at Mars. Discussions of intelligent life in the cosmos are circumscribed with fear and terror.

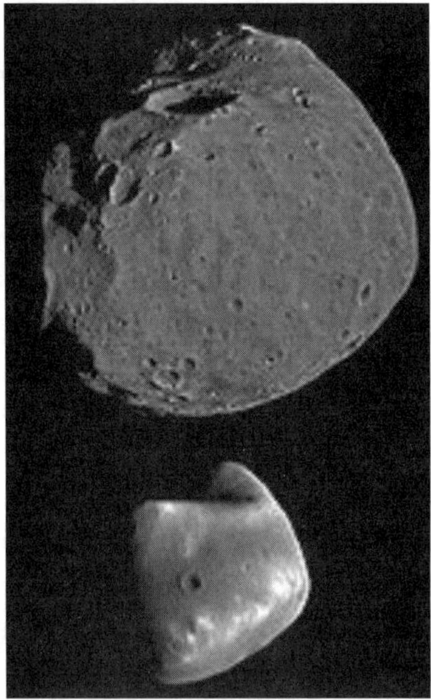

Phobos (upper image) and Deimos (lower image) (NASA)

Mars is not a planet; it is a system. Its two moons, Phobos (fear) and Deimos (terror) orbit it. One cannot look at Mars in isolation. This fact is extremely convenient for the purpose of humans setting foot on Mars's surface to confront its mysteries face to face. However, the fact that fear and terror encircle Mars is also a metaphorical truth. The fact that we may confront life and death on Mars speaks also to the human future in the cosmos: Will we be victors or victims in the larger story of a living cosmos?

Most scientists are extremely uncomfortable speaking of these matters. For one thing talking about things that provoke fear and terror is unpleasant, and secondly there are no publicly known facts relating to intelligent life in the cosmos. The worst sort of fear is fear of the unknown, partly because it deals with the unknown. Scientists often dodge questions on the subject of the fear and terror associated with ETs (Extra-Terrestrials) by implying that such questions and fears are illegitimate. There is another reason scientists don't like to discuss the moons of Mars, fear and terror, in the context of their work on Mars: Scientists get their money from the government and the government does not want this discussed. However, Tyr, the Mars of the Nordics, was also the god of the Allthing, where all things could be discussed. Therefore, we will discuss this thing.

If we have found life on Mars then, by logical extension, we have found that life and its evolved form, intelligence, is probably common in the cosmos. By the principle of Mediocrity, we are not an aberration in the cosmos; therefore, people like us are to be found elsewhere. Humans are aggressive and warlike, so might be the ETs we encounter.

In 1905 H. G. Wells published *The War of the Worlds*. It was the first book of the modern era to discuss the results of human ET contact and it articulated the unspoken fear all intelligent people held concerning the Lowellian Mars: People on Mars the planet, named after the war god, might hold no love for humanity. This was a completely legitimate concern in its day. It was the prime of the British Empire, and of conquest and colonization by the European powers. The age H. G. Wells lived in featured clashes of civilizations with vastly different levels of technology, and it showed the results of such clashes. Wells knew how humanity behaved, so he merely had the Martians behave the same way. So in contemplating a possible future confrontation with a civilization on Mars, people confronted fear and terror. It is for this reason still, that the scientific community can never approach the question of life on Mars in isolation, the moons Phobos and Diemos rule the sky above it, and behind smug skepticism, lurks the specter of denial.

In *The War of the Worlds* H.G. Wells presented the worst case scenario of contact between two civilizations in the cosmos. The novel presents a tableau of fear and terror. It was literally the Earth as Poland in 1241 with the Martians playing the role of the Mongols. However, it was a worst case scenario, and brightened at the end by evidence that Martians, while brilliant, were guilty of poor planning in not immunizing their troops against Earthly diseases. Thus, the Martians succumbed to hubris. Therefore, H.G. Wells invoked the idea of universal truth, that not only was intelligence and ruthless greed found on other worlds but so was poetic justice, for it was the simplest things of Earth that overthrew the complex schemes of Mars. The idea of universal truths that apply on all worlds, not just the laws of physics, but moral laws, is a powerful concept we must explore as we contemplate the meaning of what we may have found on Mars.

Phobos (brighter light) and Deimos (dimmer light) imaged from Mars surface by the Opportunity Rover (JPL/NASA)

Intelligence is favored in predators on this planet, and there is no reason to expect this to be different on other worlds. The killer whale is probably the second smartest creature on this planet, and it is a voracious predator. Unfortunately, any student of human history can predict, based on Mediocrity, that some species of ET may be aggressive and warlike. However, many societies on Earth have recognized the advantages of peace in their international dealings, and international organizations, both regional and global have more and more become agencies for solving disputes between nations. Therefore, while hostile and warlike species make interesting reading in science fiction, it can be reasonably hoped that most species in the cosmos have found peace with their neighbors to be far more fruitful. This is based on our experience on this planet, which under the assumption of Mediocrity can also be applied to the peoples of the cosmos.

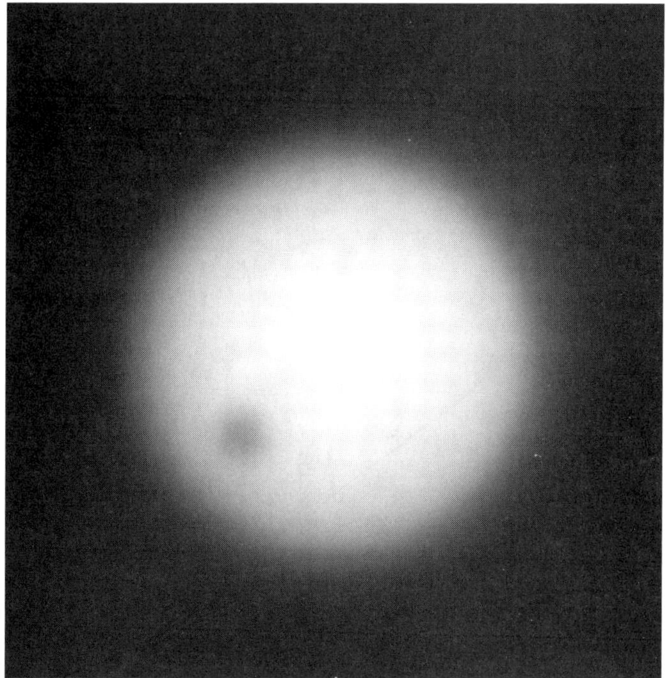

Deimos crosses the Sun's disk, viewed from Mars's surface by the Opportunity Rover (JPL/NASA)

Phobos crosses the Sun's disk, viewed from Mars surface by the Opportunity Rover (JPL/NASA)

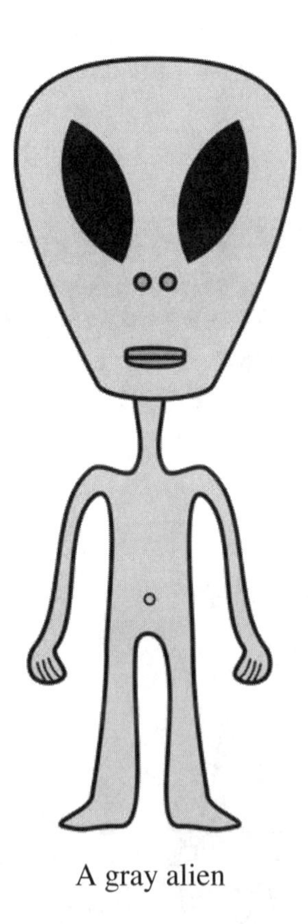

A gray alien

The WAR of the WORLDS
By H. G. Wells
Author of "Under the Knife," "The Time Machine," etc.

The War of the Worlds cover

Based on the probable finding of life on Mars, we should expect signs of intelligent life to be found elsewhere in the cosmos, perhaps a radio broadcast, perhaps even a visit. The character of the ET visit, based on Mediocrity, could go either way. However, there has been no grand visit, no noisy broadcast; we have instead silence. This prompted the great question posed by Enrico Fermi, when he asked, "Where the hell are they?" for he had already arrived at the conclusion that humanity was not alone in the universe. Therefore, if there is life on Mars and therefore elsewhere, why no great sign of it? Where are the ETs we would logically expect to see or detect if Mars holds life?

Some might suggest that UFO activity should not be considered in discussions of our place in a living cosmos. If UFO activity, at its essence, is due to ETs, then Mars science and the UFO phenomenon are sending us the same message: We dwell in a living cosmos. However, the UFO phenomenon presents its own set of problems to the scientist.

The ET is to the UFO phenomenon what the quark is to subatomic physics, something that explains everything that is seen simply, but only if the quarks themselves behave in a complex way. In particular, the quarks must obey a rule forbidding them to be seen in isolation. So the quarks explain everything and move everywhere powerfully, but they are bashful and the ETs are the same. However, unlike subatomic physics, whose theories of quark behavior can be tested and verified every day at will, UFO incidents are mostly random events and usually poor in data. Therefore, UFO activity is dismissed by many scientists since it requires ETs to behave in a complex way. It would also require the governments of Earth, particularly the superpowers, to behave in a complex way.

Orson Wells at the Mercury Theater, with headline about the terror it caused

The superpowers have a wide network of intelligence gathering apparatus, and the information from these networks flows to centralized command centers. If anyone on Earth was to know if UFO activity meant that ETs had come here it would be the superpower governments. However, after the *War of the Worlds* broadcast and the public's reaction to it, no government would be eager to share knowledge of an ET visit if it was aware of it. The War of the Worlds broadcast showed the government that news of an invasion from outer space could cause terror. The government fears terror more than any other thing. Terror means that the public panics and becomes a disorganized mob and government becomes ineffective. Terror means loss of power for the government. Therefore, terror, or its potential, must be fought with all means available. This includes withholding information from the public. For this reason, the discovery of life on Mars, and by implication, a living cosmos, is something the government will try to manage carefully. The same goes for any government detection of intelligent ET activity. So, the seeming silence of the stars is matched by the silence of the governments concerning UFO activity and its meaning.

One possible scenario of covert contact between the government of the United States and UFO aliens was explored by the author, under his science-fiction nom de plume Victor Norgarde, in the novel, *Morningstar Pass, The Collapse of the UFO Coverup*. In this novel the concept of Mediocrity is applied to the cosmos in order to weave together UFO phenomena with a scientifically plausible drama. The UFO cover-up is real in the novel, and dates from the depths of the Cold War. The UFO cover-up unravels in the novel like the Watergate and Iran-Contra-gate cover-ups, with a torrent of leaks and finally Congressional hearings. In the end, the truth of the human place in the cosmos becomes known publically, as is inevitable. It is the nature of truth to reveal itself. In any case the only reason any real UFO coverup could remain successful would be that our ET neighbors remained basically non-intrusive.

One cannot hide from the cosmos. We have been broadcasting our presence to the cosmos loudly for nearly a century. Whoever can come here in response will come. The signals cannot be called back. We are loud enough that we could detect our own broadcasts if they came from a planet 100 light years from here. This means more advanced civilizations could detect us from still further if they were looking. The further the sphere of radio broadcasts expands at the speed of light the greater the chances it will be detected by someone. Or if it has been already detected by several peoples, then the number and variety of peoples detecting it must grow with time. We should be easy to detect, and if a means is possible to cross many light years (a big if) some ETs could pay us a visit.

The full range of outcomes of an ET visitation of Earth have been explored in science fiction. These range from positive, such as ET The Extraterrestrial and The Day the Earth Stood Still, to the negative, such as Independence Day and the first ET visitation novel the *War of the Worlds*. At least most of these fictional treatments end happily, so we will consider that humanity is at least making up good stories about the future. That is a good sign. Given that we are still debating the existence of life on Mars in some circles, and by extension, in the wider cosmos, any arrival of members of our stellar community has been minimally disruptive to our civilization in its public life. If the superpower governments knew the UFOs were extraterrestrial then the ETs have certainly been very cooperative with the governments on Earth in keeping this from the public.

No, this is not a dinosaur skeleton; it belongs to second smartest predator on Earth, the Killer Whale

We can conclude from these speculations that if any neighbors in the interstellar community exist, who could come here, they have not been intrusive to any significant degree. Like the quarks, they are bashful. This suggests that the attitude of any real neighbors in the community has ranged from polite to scientifically curious. This is Deardorff's Leaky Embargo Hypothesis. However, this does not mean that the UFO phenomenon has left Earth untouched.

Cortez and his princess, impersonating Queztacoatal

The UFO phenomenon, whatever its core cause, real or imaginary, has profoundly inspired technological development on Earth and our thoughts about ETs. The symbol of the flying saucer and the "gray" type ET have permeated our culture, despite being consciously ignored by most scientists who talk about ETs. The UFO phenomenon has affected everything from how we play, the Frisbee, how we visualize our own future, the saucer-like USS Enterprise of Star Trek, to how we design buildings, large domed structures, or favoring discs at tops of towers, like the Space Needle in Seattle. It has been suggested that technology recovered from downed saucers has provided inspiration for several innovations in computers and communications. So a form of cultural transference may have already occurred, even if it was actually imaginary. Writers of the future may say of this period in human history that the cosmos spoke to humanity with silence, and being unnerved by this, the human race began hearing voices, but the voices spoke wondrous things.

Alternatively, future writers may say instead that UFO phenomena were like the methane-like absorption bands of Mars, seen in the early 1960s by Sinton, something that the-powers-that-be of the time simply did not want investigated. They may also find hilarious the fierce debates over the microfossils of ALH84001, while on the same night formations of UFOs flew over Phoenix. But the writers of the future will have the benefit of knowledge we do not have.

So, we of the present, are left again on the sands of Mars, with the silent stars offering us questions without number.

Buildings in Shanghai

Therefore, assuming life on Mars is found, and by extension a living cosmos, how does one handle the fear and terror that encircle such a discovery? What shall we do now? Might some in the cosmos be unfriendly? What will we do if ETs show up here, and tell us to turn down our stereos? If they are out there why don't they talk to us? You must answer your own questions yourself.

The first thing to realize upon finding life on Mars is that the cosmos remains as it always has been, a fairly benign place to live, and it is only we who have changed. The answer to fear is courage and the answer to terror is preparation. The school of Mars instructs us to keep a cool head, hope for the best, and prepare for the worst upon finding ourselves in a living cosmos.

Our reaction should be positive and enlightened. We should continue to invest in civilian space technology, gaining skill, expanding our knowledge and presence in the solar system that seems to have been left for us to enjoy. The Moon and Mars should be visited, outposts established there, and they should be finally colonized. The human presence in the cosmos must grow and expand.

Earth and its Moon seen from orbit around Mars (JPL/NASA)

We should also make reasonable preparations for deterrence, should our good fortune run out and someone truly hostile should appear. Remember that we learned all these things about the cosmos at Mars. Therefore, let the sword be kept in its sheath until all else fails, but bear the sheathed sword none the less, and let it be kept sharp. It will save much more heartache than it provokes. In short, since we most likely live in a living cosmos, we should resolve to move boldly outward to the stars and take our place in this living cosmos. We should be prepared for all contingencies in a rich and complex universe, where both life and death occur.

In the history of Earth, no more shining example is offered of a nation finding itself in a wider world full of more advanced nations, yet triumphing in the end, than Japan. Japan, with the help of the United States, kept its sovereignty and culture during the period of aggressive European colonization and also rapidly absorbed western technology to become a major economic power. Japan's strong military and remote location kept the European powers at bay, while its leaders pondered the best course for it to follow. It was this modernizing Japan, exotic, yet so dynamic, that inspired Lowell to turn his eyes to Mars. After some difficulties, Japan has emerged as one of the world's leading cultures. Humanity can learn much from Japan's example, both its successes and mistakes, in this new cosmos where we find ourselves.

Japanese woman writer from the late 1800s

The first Japanese Navy steamship from 1870

Percival Lowell, the revealer of both an awakening Japan and a mysterious Mars

Our reaction to finding life and death on Mars should be positive and bold. It appears Mars was the scene of massive life and massive death, perhaps even an unnatural nuclear catastrophe. We must go up to Mars and find out exactly what transpired and thus learn as much as possible about life and death in the cosmos. If we are, as some argue, largely alone in this corner of the cosmos, then we should be quick and exploit this good fortune. If we are being watched, as some others argue, let us give the observers a good show. Let us show ourselves quick.

Thus, in order to fully grasp the meaning of finding life on Mars, you must conquer the moons of Mars, Fear and Terror, that encircle it. You cannot deny these moons exist; you must instead deal with them. You have graduated from the school of Mars; put this education to good use. The school of Mars instructs us that one must recognize good fortune and take advantage of it.

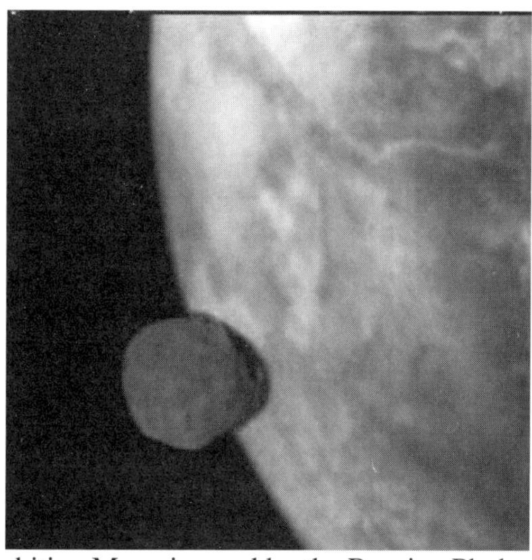

Phobos orbiting Mars, imaged by the Russian Phobos mission (NASA)

We face the Second Copernican Revolution. Whereas before, we let go of the concept that we were the geometric center of the Cosmos, and prospered, now we must let go of the concept that we are the living center of a dead universe. We are not alone, you must face up to this brave new cosmos.

Again, Mars is not a planet; it is a system. Deimos and Phobos orbit it. In a purely practical sense, they make fine places to land. They are the easiest places to go after the Moon. People complain that Mars has no significant magnetic field like the Earth, and so has more cosmic rays on its surface. But since it has no magnetic field it also has no Van Allen belts, making a human landing on Phobos or Deimos trivial. One does not land on these moons so much as dock with them. Once you have conquered the moons Fear and Terror, the surface of Mars is but a short step away. The moons make excellent bases in the Mars system, being ready-made space stations. Phobos is 22 km across and orbits a mere 6000 km from the Mars surface. From its surface Mars fills the sky. So you can conquer fear and terror, and only then can we deal with the wide cosmos as it is, living, vast, complex, full of dangers, opportunities, and with few givens. Boldly go forth.

The Epilogue of Mars

"Everything I needed to know about life in the Cosmos I learned at Mars"

Human voice of the future

So we have received a message in a rock, hurled from Mars, that killed a dog. It was payback on the canine species by old Tyr, but it is also a message to the wise. It was a message of life and death. The rock that killed, Nakhla, carried a message of life inside it. The message is vitally important. To humanity, the New Mars Synthesis says that we are approaching a turning point as an intelligent species in the cosmos. This is a happy day. It is hoped that future generations will conclude that the planetary science community of our generation was on the whole, learned, and in the endgame, quick.

The generation of scientists who ran Martian affairs up until 2000 were the Young Turks of the 1960s who overthrew the Mars of Lowell. In their day they were heroic, fearless, and hard eyed. When the canals disappeared from the maps of Mars and were replaced with a lunar landscape they laughed in triumph. Dead rock was their model of Mars. Lifelessness was the entity that explained all things. The Lunar Mars model, which was their banner, swept all before it. To them biology was complex, wet, Byzantine, and was never the simplest hypothesis. It had no place in Mars science. Now it is apparent that Mars is not the Moon. Mars is a unique entity. Now this "Young Turk" generation confronts a wall of data that cries out "life." Life has become the one entity that simplifies everything we see on Mars. It is understandable that many present day Mars scientists resist this. They, in fact, invoke the most byzantine of scenarios of lifelessness to resist the simplicity of life, crying out that life is complex and therefore never the simplest hypothesis. However, the "simplest hypothesis" can never explain what is seen every day on Earth, even if the concept of "a minimum of entities" can. This is a new age, and it requires new ways of thinking. How much better for us humans, if that new way of thinking is merely old wisdom, understood in a new way.

A new generation of Young Turks has arisen on Mars, and they are moving swiftly to bring this long search for life to a conclusion. The Mariner probes that found water channels were followed by the Vikings. Mariner 9 found evidence for an early warm-wet Mars, but this assumed to be too transitory to have allowed life to develop. The Vikings probably found life and led to the search for more evidence of liquid water. The Mars meteorites were found and confirmed both life and liquid water in Mars's past. The search for water led to the search for hematite and once that was seen from orbit a rover was landed right on top of it to confirm it as a water-formed sedimentary mineral. Then the methane vents were found nearby. With no geologic hot spots in evidence, the simplest remaining explanation for the methane plumes is anaerobic bacteria feasting on buried organic matter from some previous epoch of abundant life. Now we are preparing a nuclear-powered rover to investigate the methane plumes.

We are a fortunate people in the cosmos. If we were not we would not be having this conversation. Our luck as a species has allowed us to escape many of the dangers present in the cosmos, asteroid impacts, nearby supernova explosions, gamma ray bursts and hazards that we cannot yet understand. The cosmos has provided us with a beautiful planet that continually gives us everything we need. We need to take care of it, but we need to do more than this. The fact that an unprotected human being cannot survive for even a minute on the surface of any other planet or Moon in this solar system hints at a harsh truth of the cosmos, that the cosmos gives death as well as life. Not believing in asteroid impacts will not give protection from them, or any of the other hazards of living in the cosmos. The unexamined life is not only not worth living, in the words of the ancient Greeks, but, in the cosmos as we now know it, it is also short. We could wake up tomorrow, for example, to the news that a large asteroid has been discovered and is found to be on a collision course with Earth. If that happens, it will be the lance of Mars, our space technology, our knowledge of nuclear weapons, and the power of prayer, that may be all that stands between us and extinction. The cosmos is a battlefield of life and death; therefore, be prepared. Be quick.

The cosmos is a deadly desert and we live on an isolated oasis in it. However, this realization is not an admonition to remain on our oasis. It tells us we must journey out across the desert and learn its ways, for the oasis we live on is a fragile and precious home in a larger and harsher world. We must take up the challenge this cosmos presents to us and boldly go forth to meet it.

Mars, both the physical planet and the evolving concept that it has presented to humanity from ancient times, has been a teacher to us. You cannot deny Mars, it is part of us. If Mars is telling us something about the cosmos, you ignore it to your peril. This book is an attempt to decipher that message that Mars has sent to us, as we understand it now. Mars tells us this message in the medium of cutting-edge science as well as ancient folklore. The message Mars has sent us, in brief: the cosmos is a place of both life and death, of nurturing abodes of life and almost unimaginable catastrophes that erase them, Mars's first words of this lesson are *"do not dwell carelessly ... be quick."* May you and your children learn well in the school of Mars and live. Choose life.

Thus the tale of two planets can be discerned now, a tale drawn from the accounts of the stargazers and peerers through telescopes, of dreamers, of gatherers of meteorites, of builders of rockets and finally builders of spacecraft and rovers. It is the tale of the Odyssey, Vikings, and the Mariners, the Surveyors and Rovers, and Mars Express. Finally Mars is close to yielding its great secret, and with it the secret of humanity's place in the cosmos.

The search for life on Mars has come to resemble a Victorian romance, full of passionate letters, smoldering looks, and great sighs, but alas, no consummation. Curmudgeons have been the leaders of these debates rather than Young Turks. Many scientists have confessed to me that finding evidence of life on Mars is an occasion for fear rather than joy, because the professional hostility such a finding attracts is so severe.

The Opportunity in Terra Medianis near Arabia Terra (NASA)

But the human race persisted and investigated. Mars rose Phoenix-like from death after Mariner 4 to be a focal point of controversy, rumor, and intrigue. Those who said Mars was as lifeless as the Moon held sway for a while, and focused on the rocks of Mars. However, the rocks spoke and said that Mars was much more like the Earth than the Moon. Finally the rocks cried out that Mars had nurtured life. Now the search for life on Mars is hot and in its end game.

Nozomi (Hope) spacecraft launched by Japan (JAXA)

Finding life on Mars, even fossils of bacteria, will be a triumph of science, but it is also like finding a rotting corpse in your backyard. That is the stark truth. H. G. Wells spoke to a real fear when he had Mars brutally invade Earth in fiction. Wells grasped an essential truth about a living cosmos that people remain uncomfortable with; he understood that such a living cosmos would be dangerous. Mars truly has two moons named fear and terror. They are small but they are there. Mars is not a planet, it is system.

The discovery of biology on Mars means not only life, but death, are close by us. Events of almost unimaginable violence and mass death, perhaps even malice, may have occurred there. Such possible scales of malice would be considered impossible, if we had not seen them on Earth. We must go and discover what really happened on Mars. And we should be quick to do this.

In the 1960s movie, *Robinson Crusoe on Mars*, an astronaut is stranded on Mars with only a monkey for company and manages to survive by finding oxygen in the rocks and underground water. He struggles to survive, and finally succeeds. To celebrate his mastery of the art of living on Mars, he decides to explore his harsh new home. In an epic scene the astronaut marches boldly across the plains of Mars with his rudimentary breathing apparatus, happily blowing a makeshift bagpipe for the benefit of his pet monkey, but he then suddenly encounters a gravestone, and a humanoid skeleton buried beneath it. In one instant, his feelings of levity and loneliness depart and are replaced with fear and watchfulness. The skull of the skeleton bears unmistakable signs of violence. He retreats to his cave and erases all outward signs of his presence. The discovery of life on Mars means life is close, and so is death. If we are not careful, and not prepared, the life of this species may be cut short. We must "step lively."

Some would react to this finding by arguing a retreat from space, that only robots should leave Earth to explore the cosmos. This is like a man who decides to live forever in the basement of his mother's home, and surf the net. He twitters to his friends of visiting Paris, yet he has never climbed the Eiffel Tower and felt the winds caress his cheeks, he has only looked at images of its vistas. He meets a girl on the net, has virtual dates with her, but alas, their love is never consummated. They never even kiss. In the end, his agoraphobia is complete and he fears even to surf the net. Finally the house he lives in is condemned for a new freeway overpass, and he is led from it old, pale, and bewildered, his virtual world shattered by the realities of the real dynamic universe that existed around him. He declines rapidly and dies, not of broken heart, but of astonishment. A friend remarks sadly, after his funeral, "he was intelligent, but he was not quick."

We must go into space because we are already in space, we ride on a great spaceship called Earth. The world is no longer the Earth, it is the cosmos, we must venture forth boldly and explore it. We can learn everything about Mars with robots now, except how we will go there ourselves. Again, the broad cosmos is now our world, and we must grow into it to our full potential as a people.

Finding life on Mars will mean that we have explored the first terrestrial planet we can get to in the cosmos, and it was alive. Taken with the record of life, and mass extinctions, on Earth, it says that we live in a living, rugged cosmos. More planets are known to orbit other stars now than to orbit our own. The cosmos is full of planets like Earth and Mars, and finding life on Mars means some of those planets are also alive. Finding life on Mars means that when we look at the stars at night many are looking back at us. Perhaps some look with envious eyes. Denying harsh truths does not make them go away. It is a proverb of the school of Mars that "the harsh truths of life are the first truths one must consider." That is why Tyr was god of the Allthing. Harsh truths require things of us, in our thoughts and actions, and the first thing it requires is discussion.

The discovery of life on Mars, and hence life throughout the cosmos, will require a revolution in our thinking and actions. This long Victorian romance, the search for life on Mars, is going to end, as it should, with a rather lusty night in a hotel. We have chosen life. Mars had life and probably still has it. The glory of intelligence is not the recognition of an isolated fact, but in recognition of patterns. The simplest explanation for the totality of things we know about Mars, its redness, the trace of oxygen in its atmosphere, the contents of its meteorites, its old water channels, a young ocean basin, the positive Viking life tests, and its methane plumes, is that Mars has life and used to have much more of it. Mars was the scene of both massive life and death.

This is the basic story of Mars: Mars and Earth began similarly, with oceans and warm surface temperatures. We know life quickly began on Earth; we have no reason to believe otherwise about Mars. A meteorite sampling this early environment of Mars confirms its warmth and wetness and bears signs of life. The geothermal enhanced greenhouse existed on early Earth and faded only with the triumph of photosynthetic life; this same greenhouse regime persisted on Venus and also on Mars. The fact that the youngest parts of Mars bear a fossil ocean basin and river channels requires and confirms a persistent greenhouse. The high oxidation state of younger Mars meteorites and the redness of deep sediments exposed in Vallis Marineris canyon confirm the mechanism for the chemical stability of this CO_2 (and perhaps methane) greenhouse: oxygen. The absence of carbonates and the dominance of sulfates and phosphates in Mars's soil echoes this evidence. A vast and powerful geochemical engine operated on Mars to preserve its greenhouse by maintaining an acid environment and pumping back CO_2 into the atmosphere. That vast and powerful geochemical engine was life. Through life, photosynthetic plants created oxygen to promote an acid ocean and bound CO_2. Other Mars life ate the plants and exhaled the CO_2. Thus were the days of the Martian Gaia. This condition persisted on Mars for close to 4.0 billion years, based on the ages of the meteorites that contain traces of life. These include the fateful meteorite Nakhla, which is 1.3 billion years old, and brought news of life and death in the same impact. The Mars Gaia was then cut short by a massive asteroid impact that formed the Lyot basin. Thus, Mars lived for 4 eons and died in one day. Be quick, my friends, or Earth will share Mars's fate.

A faint remnant of Mars's past biosphere survives to this day, and some of its anaerobic portion feasts on the buried organics of its past glory. The methane plumes are the fumes of biologic decay of a massive dead biosphere. These plumes were ignored before, but they are now in the crosshairs. This basic story is the essence of the New Mars Synthesis. One may quibble with this or that piece of evidence, but the picture these pieces form is unmistakable. Mars holds a record of life and death of a planet. Finally, and mysteriously, this record includes a bizarre nuclear event after the Lyot event. Whether this terrible event was due to a shift in ground water and some buried thorium and uranium rich body, or something else, we cannot tell. The massive event left no visible crater. We must go and investigate, as human beings, so we don't end the same way. If anything is worse than being demoted to living on a speck of dust in a vast universe full of other living specks of dust, it is becoming a dead speck of dust. We are not alone; deal with it.

It can be said that the human race needed a shock absorber for its psyche in all of this, some delay in the recognition that they were not alone, to give everyone time to adjust. Governments, ever mindful of the legacy of the Wells family, may have contributed to this. They just didn't want people getting worried about the sky falling on them one day.

However, it can also be said that Mars has provided a cosmic intelligence test for humanity, and that we have finally passed it. The dried rivers of Mars, once flowing with water, were first seen by Mariner 9 in 1972. Now we have finally digested their meaning; it has taken 40 years. Life on Mars did not hide from us, it basically is venting clouds of methane to announce its presence. It flung itself here on rocks. Perhaps we were not ready then, but we are ready now. However, we should not go slowly anymore, the cosmos we are in is a living cosmos, and we must move out and engage it. This cosmos contains the quick and the dead. Therefore, be quick. We have strong evidence of life, we should act on it.

We should vigorously go to Mars and occupy it. This should be the focus of our space program. Waves of robotic probes should prepare the way. However, until humans set foot on the planet's surface, we have not really gone there. Once we arrive we should stay and build a human community there. As the astronaut Harrison Schmidt, one of the Americans who first stood on the Moon, has said, "Until humans live and die in a place, they have not really been there."

Mars is not the Moon, to be reached in two days and returned from within a week. A mission to Mars is a journey and sojourn of years. LOX-Hydrogen rocket fuel, which was sufficient for the lunar journey, is inadequate for Mars; its rocket exhaust is too cool and slow. To get to Mars we will use rocket engines that use plasma as hot as the Sun's surface. The author is currently involved in research to speed our astronauts to Mars using megawatt solar power arrays powering and advanced MET (Microwave Electro-Thermal) engines using water vapor as propellant. Mars has water, we can thus refuel there for the journey home.

An MET electric rocket engine being tested for Mars Mission propulsion
(Courtesy Orbital Technologies Corporation))

The planet Mars has enticed us through the ages, giving birth to both the science and technology that we have now used to investigate it and discover its secrets. When humans set foot on Mars, it will be the consummation of a great cycle of human thought and yearning that has stretched back through the ages. Once, we thought we would like to go to Mars, then we realized we could go to Mars, now we must go to Mars.

So what is the sum of these matters? It is this: The cosmos is a place full of life, but as Mars also teaches us, it is also a place of death. Mars instructs us: "Dwell not carelessly in this new cosmos you have discovered O child of Earth. Choose life. Keep your eyes sharp, and your sword as well; be strong and quick." What does this new knowledge require of us? It tells us to hold the life of our people as precious and quickly move to safeguard it. Let us go up to Mars and find out exactly what transpired there, in person. Do not pretend that visiting Paris on the web can replace a real night there. We have left our mother's house and now walk the wide cosmos. We will go to Mars and walk there, and occupy it. There may indeed be those who already gaze upon this Earth with envious eyes. Good, let them send back reports to their home worlds, that the children of Earth are graduates of the school of Mars. This will give weight to the counsel of the wise on these far away home worlds, that the children of Earth should not be trifled with. The human place in the cosmos will be two planets now. Mars is the gateway to the stars. Therefore, God-willing, the stars themselves will someday become homes to us. But only if we quickly connect the dots.

The best safeguard for a long good life for our species is that we should be expansive, vigorous, alert, and ready for any encounter or opportunity that is offered in the cosmos. Let us move out into the cosmos. Let us show ourselves quick. Let it be reported that the humans dream boldly of flights across the stars, and move quickly to accomplish this. Know that Mars is now much closer than it was in the days of Goddard.

Robert Zubrin, president of the Mars Society, in a major human intellectual advance, has showed us how this voyage to Mars shall be accomplished at minimal cost, by using Martian resources. Mars is a place of plenty to those who know how to harvest its bounty. Mars is a fuel dump and a rich farm to those who go there properly prepared. In particular, the water and air of Mars can make rocket fuel for the return trip to Earth. The carbon dioxide and nitrogen of the Mars atmosphere can, once compressed to near the Earth's ambient conditions, grow food and make oxygen for breathing. Mars is a place to live for those who are smart and diligent. These conceptual advances make Mars much easier to journey to and return from than before. Also, it is much easier to live on. Therefore, let us go out to Mars. Let us return to the planet of our intellectual and perhaps biological heritage, and restore it to life. Let the great cycle of thought and life and death be brought full circle.

Humans land on Mars, the great cycle is finished

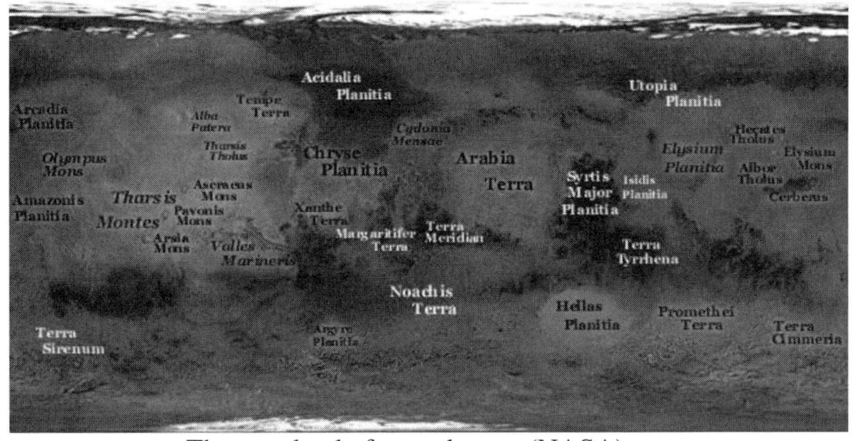

The new land of your dreams (NASA)

So both Mars and Wisdom would say to us, *"In this cosmos, you must be vigorous, imaginative, daring, and attentive, and you must be quick."* You must choose life, and cherish it. You must rush out into the starry gulf yourselves. Live, flourish, find your destiny in the community of peoples of the stars. But, wherever you go in the cosmos, remember the school of Mars. In this cosmos are the quick and the dead; therefore, be quick.

"I call heaven and Earth to record this day against you, that I have set before you life and death … . Therefore, choose life, that you and your children may live."

Deuteronomy 30:19

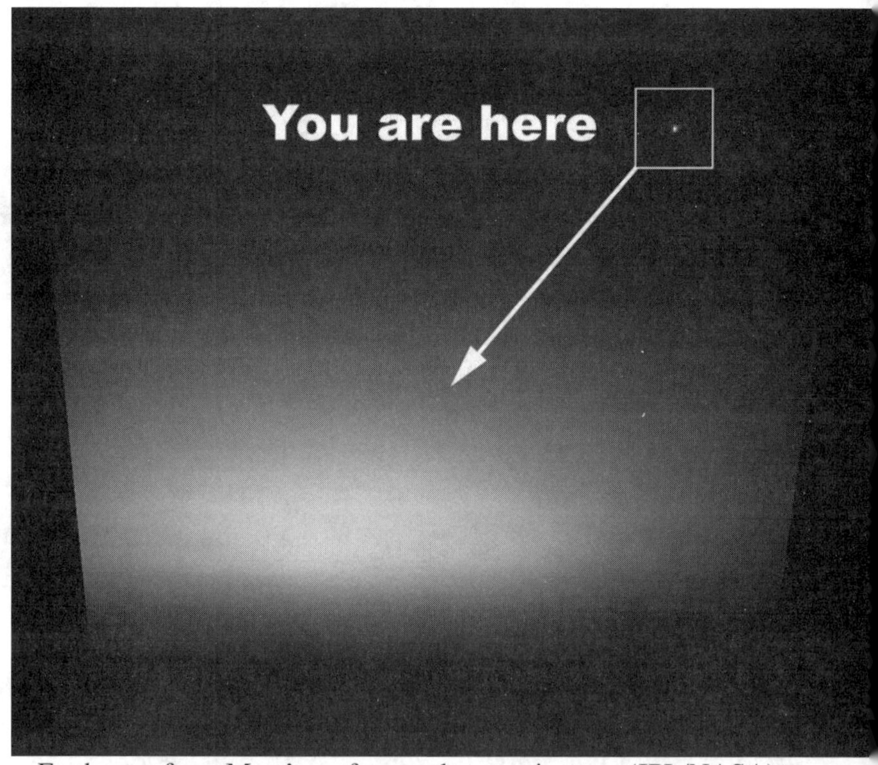

Earth seen from Mars's surface as the morning star (JPL/NASA)

Chapter Bibliography

Foreword. Oasis Earth

Sagan, Carl. Cosmos, New York: Ballantine Books, 1985.

Wallbank, Walter, Alastair Taylor, and Nels Bailkey. Civilization, Past and Present. 3rd editon. Scott Foresman and Company, 1967.

Chapter 1. The School of Mars

Gardiner, Sir Alan H. Egypt of the Pharaohs. Oxford University Press, 1961.

Wallbank, Taylor, and Bailkey. Civilization, Past and Present. 3rd editon. Scott Foresman and Company, 1967.

Chapter 2. The Dream of Mars

Sagan, Carl. Cosmos, New York: Ballantine Books, 1985.

Stavrianos, L.S. The World Since 1500: A Global History. 2nd Edition. New Jersey: Prentice-Hall, 1971.

Wallbank, Taylor, and Bailkey. Civilization, Past and Present. 3rd Edition, Scott, Foresman and Company, 1967.

Chapter 3. The Red Star

Chandler, David. Life on Mars. Clarke Irwin and Company, 1979.

Sagan, Carl. Cosmos. New York: Ballantine Books, 1985.

Stavrianos, L.S., The World Since 1500: A Global History. 2nd Edition. New Jersey: Prentice-Hall, 1971.

Carl Sagan, and Jonathan Norton Leonard, Planets,Time-Life Books New York NY, 1966

Maguire, W.C. (1977) Martian isotopic ratios and upper limits for possible minor constituents as derived from Mariner 9 infrared spectrometer data. Icarus 32, 85-97

Chapter 4. The Vikings of Mars

Brandenburg, J.E. “Constraints on the Martian Cratering Rate Based on the SNC Meteorites and Implications for Mars Climatic History.” Earth Moon and Planets 67 (1995): 35-45.

Chandler, David. Life on Mars. Clarke Irwin and Company, 1979.

Sagan, Carl. Cosmos, New York: Ballantine Books, 1985.

Chapter 5. The Oxygen of Mars

Brandenburg, J.E. "Mars as the Parent Body of the CI Carbonaceous Chondrites," Geophysical Research Letters, Vol. 23 (1996): 9, 961-964.

Chapter 6. The Paleo-Ocean of Mars

Brandenburg, J.E. "Constraints on the Martian Cratering Rate Based on the SNC Meteorites and Implications for Mars Climatic History," Earth Moon and Planets 67 (1995): p35-45.

Brandenburg, J.E. "The Paleo-Ocean of Mars" MECA Symposium LPI Tech. Report 87-01 (1986): 21.

Chandler, David. Life on Mars. Clarke Irwin and Company, 1979.

Hanlon, Michael. The Real Mars. Basic Books. 2004.

Head, James W., et. al. "Large Standing Bodies of Water in the Past History of Mars: Further Tests for their Presence using MOLA data." AGU spring meeting 1999.

Nyquist, L.E., L.E. Borg, and C.Y. Shih. "Shergottite Age Paradox and the Relative Probabilities of Martian Meteorites of Differing Ages" JGR (Planets) 103 (1998): E13, 431-445.

Parker, Timothy et al. "Geomorphic Evidence for Ancient Seas on Mars" MECA Symposium LPI Tech. Report 87-01 (1986), 97.

Chapter 7. The Crystal Palace of Mars

Sagan, C., O.B. Toon, and P.J. Gierasch. "Climate Change on Mars." Science 181 (1973): 1048.

Chapter 8. The Chicxulub of Mars

Brandenburg, J.E. "The Big Chill: Did the Lyot Impact produce a late climate Cooling On Mars?" 2002 Meteoritics Society meeting Los Angeles. 2002.

Norgarde, Victor. Asteroid 20-2012 Sepulveda. 2009.

Russell, P.S. , and J.W. head (2002) The Martian Hydrosphere/Cryosphere System: Implications of the Absence of Hydrologic Activity at Lyot Crater, GeoPhys. Res. Lett 29(17),1827

Chapter 9. The New Mars Synthesis

Hanlon, Michael. The Real Mars. New Jersey: Basic Books. 2004.

Chapter 10. The Twilight of Mars

Brandenburg, J. E. "Evidence for a Large Natural Nuclear Reactor in Mars Past." Proceedings of STAIF 2006. Albuquerque, N.M.

Brandenburg, J.E. "Evidence for a Large Natural Nuclear Reactor in Mars Past" Proceedings of Spring AGU meeting, 2006.

Chapter 11. The Endgame of Mars

Hanlon, Michael. The Real Mars. New Jersey: Basic Books. 2004.

Chapter 12. The Moons of Mars

Norgarde, Victor. Morningstar Pass, The Collapse of the UFO Coverup. 1st Books Library. 2003.

Sagan, C., and I.S. Shklovskii. Intelligent Life in the Universe. New York: Random House. 1966.

The Epilogue of Mars

Zubrin, Robert. The Case for Mars. Free Press. 1996.

Index

Evidence for a Large Natural Nuclear Reactor in Mars Past

J.E Brandenburg

Florida Space Institute
University of Central Florida
MS:FSI
Kennedy Space Center, 32899

Abstract : The high concentration of ^{129}Xe in the Martian atmosphere, the evidence from ^{80}Kr abundance of intense neutron radiation of the Martian surface and the detected excess abundance of Uranium and Thorium on Mars surface, relative to Mars meteorites, is modeled as due to a large natural nuclear reactor that went into a fast neutron mode and exploded catastrophically, spreading residues over the planet's surface. This The reactor is modeled reactor as a kilometer scale concentrated ore body of Uranium and Thorium oxides that being infiltrated with water, went critical, and operated as a thermal neutron reactor for a considerable period, similar to fossil reactors found at Oklo in Africa. This operation bred both ^{239}Pu from ^{238}U and ^{233}U from the ^{232}Th before suddenly going into an uncontrolled fast neutron reaction because of greater neutron multiplication. Debris patterns are seen in maps of surface abundances of long-live isotopes of Th and K radiating from a site in Northwest Mare Acidalium at approximately 30 W 50 N. An approximately 10^{25} J energy release is calculated, similar to the Chicxulub event on Earth. Delayed neutrons from the global debris layer irradiated much of Mars surface to the 10^{14}/cm^2 flux required to account for ^{80}Kr abundance.

Introduction

A signature feature of Mars atmosphere is the predominance ^{129}Xe over its other isotopes(1) relative to Earth and other inventories(see Figure 1). This feature, along with ^{40}Ar, allowed the identification of Mars as the parent body of the SNC meteorites. A predominance of ^{129}Xe is also present in a component of the Earth's atmosphere due to fission chain reactions, both natural and artificial, on the Earth (see Figure 1). The contribution of ^{129}Xe is due to the shift to lower atomic isotopes to the Atomic number -129 channel, chiefly to the parent of ^{129}Xe : ^{129}I with half life of 15.7 Myr. Therefore, the signature ^{129}Xe predominance of Mars can be explained as due to both fast neutron fission of $^{235}U$and fission of ^{239}Pu and also ^{233}U, which shares this same property in fast neutron fission (2). Concentrated ore bodies on Earth have hosted natural nuclear reactors that have run cyclically for millions of years (3), however, these reactors are modeled as operating predominately on ^{235}U fission with thermal neutrons using water as a moderator. Therefore, it seems plausible that natural nuclear reactors could also operate on Mars, with a faster neutron spectrum (4, 5). Other evidence supports the idea of natural fission reactors operating on Mars.

It has been noted that the relative abundance of ^{80}Kr can be explained by neutron capture on ^{80}Br (1) requiring a neutron irradiation of Mars rock of approximately $10^{14}/cm^2$-$10^{15}/cm^2$, depending on the neutron energy spectrum. In the Shergottite EETA 79001,a composite of three distinct lithogies of approximately the same age, some lithogies show direct evidence of such irradiation (6). The difference in irradiation in lithogies of approximately the same age in the same meteorite suggests that this irradiation was a concentrated event in geologic time. The radiometric age of the lithogies bearing evidence of irradiation is approximately

180 Myr, this is long enough ago that any ^{129}I produced would decay into ^{129}Xe. It seems possible, therefore, that if natural nuclear reactors operated on Mars and were widespread, they could explain the ^{129}Xe in the atmosphere and neutron exposure of some rock samples and the abundance of ^{80}Kr in the atmosphere.

A problem with this model of widespread natural nuclear reactors on Mars is that meteoritic samples of Mars rock are mostly depleted in Uranium and Thorium relative to Earth rock suggesting that any natural reactors would have to be contained in a concentrated ore body. Support for this is found in the Phobos and Mars probe data,(7) now supported by the Odyssey GRS (Gamma Ray Spectrometer) (8), which shows enhanced levels of Uranium and Thorium, in approximately Mars rock ratio to each other, in the top meters of the Mars surface that can be accessed from orbit. It therefore seems possible that a large concentrated Uranium and Thorium ore body existed on Mars, perhaps as the result of a previous impact of an enriched bolide, and that this ore body not only hosted a nuclear reactor of large size, but it went unstable and exploded, giving rise to a global layer of debris enriched in Uranium and Thorium. Enhancement of K and Th can occur also in lavas, as is seen in KREEP lavas on the Moon (9) Thorium maps of Mars show a large concentration of Thorium in northwest Acidalia Planitia (8) (See Figure 2) with an another enhancement at the approximate location of the antipode of the principal hot spot, as would be expected from a global dispersal of debris. A similar pattern exists in maps of radioactive potassium on Mars.

Two main modes nuclear reactor operation are possible: 1. A thermal mode, ^{235}U is the primary fissioning species at low concentration and water or some other moderator is necessary to slow down the neutrons to allow criticality. This reactor can be self regulating because at high temperatures the water can boil out and the reactor loses criticality until it cools. This mode of operation appears to have operated at Oklo (3). Such a reactor can breed ^{239}Pu from the ^{238}U or ^{233}U from ^{232} Th. 2. A non thermal or fast breeder mode, where higher concentrations of fissionable materials sustain a fission chain reaction using fast neutrons without moderator. Such a reactor often operates partially on ^{239}Pu or ^{233}U and breeds its own fuel. The fast neutron even fissions ^{238}U or ^{232} Th with its high energy neutrons. In addition the fast neutrons create more neutron multiplication than slow neutrons so it is possible to have strong increases in criticality during reactor operation as the more easily fissionable species are bred and the neutron spectrum becomes faster. A fast neutron reactor does not need a moderator and hence can go catastrophically unstable and have an explosive disassembly. It can be easily imagined that a reactor could operate in a large ore body in a thermal mode and then, if the reactor bred its own fuel, transitioned into a fast neutron reactor and went unstable. Accordingly a detailed hypothesis can be formed based on these salient features of reactor physics and Mars data.

The Hypothesis

In the north of Mars, in Mare Acidalium,a large region of concentrated uranium and thorium ore was formed, probably by a past asteroidal impact, that supported nuclear fission reactions based on a thermal mode at many locations. This process began 2 billion years ago when ^{235}U was three percent. This ore body was of similar high concentration to the Oklo deposit, being almost pure Uranium and Thorium oxides. After many millions of years in operation it managed to begin breeding fuel in the form of ^{233} U and ^{239}Pu faster than it was burned up. At some point the ore body suffered a "prompt critical" and the water boiled out making the neutron spectrum harder and a runaway chain reaction ensued. Because of the size of the ore body, and it possible deep burial, the reaction was inertially confined or "tamped" so that disassembly was delayed until a high degree of fission burnup was achieved. The resulting energy release was catastrophic and resulted in an explosive dispersal of the ore body as a dust and ash cloud similar to a large asteroid impact. Turbulent mixing of the concentrated ore residues with dust and ash from the surrounding rock in the explosive plume resulted in dust and rock fall over a large areas of the planet, this layer was enriched in U and Th over the base rocks of the Mars surface. Delayed neutrons, of approximately 1% of the core neutron flux irradiated the planets surface for several minuets as debris rained down to form a global layer. ^{129}I was made in vast amounts and decayed to form the ^{129}Xe seen in the atmosphere.

Fission Yield Calculations

Based on the observed abundances of Mars Xe and Kr isotopes and the observed enriched layer of U and Thorium on its surface, it is possible to estimate the number of fissions that occurred under this hypothesis and thus the energy release and approximate size of the original concentrated ore body.

Based on the abundance of ^{129}Xe in the Mars atmosphere and assuming it was all produced in the explosion at approximately a fraction mass yield into the atomic mass 129 channel of F_{129}=3% for a fast neutron spectrum(2) we can write for the total energy released based on ^{129}Xe:

$$W_{Xe} = W_{fission} n_{Xe129} AH / F_{129} \cong 1.5x10^{25} J \qquad (1)$$

where $W_{fission}$ is the energy released per fission of 200Mev or 3.2x 10^{-11}J, n_{Xe129} =9x10^{10} cm^{-3} is the number density of ^{129}Xe in the Mars atmosphere, A is the surface area of Mars of 1.4 x 10^{18}cm^2 and H =1.1 x 10^{6} cm is the Martian atmosphere scale height. This is a large energy, equivalent to the impact of a 70km diameter asteroid into Mars and sufficient to produce a global ejecta layer of 4meters (10).

Based on the neutron fluence $F_{neutron}$ = 10^{14}/cm^2 neutron fluence required to explain the irradiation of lithogies B, C of EETA79001 and account for the ^{80}Kr anomaly, and assuming this was a planet-wide occurrence from delayed neutrons of an approximate fraction $F_{delayed}$ = 0.1% that were radiated immediately after the event by fission fragments in the planet-wide ejecta layer, we can calculate and approximate number of fissions in the event and thus have an independent estimate of the yield. We can estimate the yield from the ^{80}Kr anomaly:

$$W_{Kr} = W_{fission} F_{neutron} A / F_{delayed} \cong 4.6x10^{25} J \qquad (2)$$

where the values of other quantities $W_{fission}$ and A are the same as in Eq. 1.

Assuming a thickness L=1 meter layer of Th and U of concentration C = 1 ppm of a total molecular number density of n=6x10^{22} cm^3 covering the planet's surface and, similar to Oklo, that this is the remnants of a concentrated ore body where approximately a fraction $F_{fissionable}$= 3% of the ore body was fissionable and was consumed in the explosion, we can again estimate the total energy yield:

$$W_{U-Th} = W_{fission} F_{fissionable} CnAL \cong 8.3x10^{24} J \qquad (3)$$

The original ore body, if it was approximately pure (Oklo was 70%), would have been approximately the volume of 0.14 cubic kilometer and the explosion would have been a planetary scale catastrophe, creating a crater approximately 200 kilometer wide and 3 kilometers deep. The observed region of concentrated Th is located in Northwest Mare Acidalia centered at approximately 30W and 50 N and is a darker ring shaped area inside a major dark albedo feature on Mars. This suggests that the crater was buried by subsequent mud flows. The appearance of a region of enhanced Th and radioactive K is not reflected in maps of shorter lived Fe and Si isotopes and indicates the event occurred several million years ago and probably dates to the middle or late Amazonian epochs. Irradiation of lithogies in ETA79001 indicate a possible 180 million year age for the event.

Summary and Discussion

The evidence for a large fission event in Mars past is present in the xenon spectrum, especially the ^{129}Xe abundance. Such fission requires concentrated ore bodies. The enhancement of Thorium and uranium in a surface layer suggests such ore body existed and dispersed explosively over the planets surface, consistent with the large energy release indicated by the xenon abundance. The event was apparently recent in Mars history, in the late or middle Amazonian epochs, and was probably triggered by a change in groundwater distribution related to a global cooling event. It is possible that radiation nourished bacteria (11) may have played a role in this event by concentrating Uranium and Thorium for their own use. Delayed neutrons from the dispersed debris layer would irradiate the surface and account for the evidence such of irradiation seen in ^{80}Kr and in the SNCs themselves. Approximate calculations appear to give a fission yield of 10^{25} J. This suggests a truly catastrophic release of energy, equal to the impact of a 70km diameter asteroid, with the ability to change Mars climate.

Aknowledgments

The author is grateful for many helpful discussions with Paul Lowman of Goddard Space Flight Center and J. Marvin Herndon. The author is especially grateful to Edward McCollough for his pointing out of the antipode feature in both Th and K distributions of possible debris.

References

1. Swindle, T. D., Caffee, M. W., and Hohenberg, C. M., (1986) "Xenon and Other Noble Gases in Shergottites" Geochimica et Cosmochimica Acta, 50, pp 1001-1015.
2. Vandenbosch, R., and Huizenga J.R., (1973) Nuclear Fission, Academic Press Inc. NY NY p 307-315.
3. Meshik, A. P., Hohenberg C.M., and Pravdivtseva O. V. (2004) “Record of Cycling Operation of the Natural Nuclear Reactor in the Oklo/Okelobondo Area in Gabon” Phys. Rev. Lett., 93, 182302.
4. Brandenburg J.E. “Evidence for a Large Natural Nuclear Reactor in Mars Past” Proceedings of STAIF 2006 Albuquerque N.M.
5. Brandenburg J.E. “Evidence for a Large Natural Nuclear Reactor in Mars Past” Proceedings of Spring AGU meeting 2006

6. Rajan R.S., Lugmair G. Tamhane A. S. and Poupeau G. (1986)" Nuclear Tracks, Sm Isotopes, and Neutron Capture Effects in the Elepahnt Moranine Shergottite. Geochem. Cosmochim. Acta. Vol. 50, p 1234.

7. Surkpov Y.A., et al. (1988) "Determination of the elemental composition of Martian rocks from Phobos 2" Nature Vol. 341 p595

8. Taylor G.J. et al. (2003) "Igneous and Aqueous Processes on Mars : Evidence From Measurements of K and Th by the Mars Odyssey Gamma Ray Spectrometer" Proc. 6th International Conference on Mars.

9. R. C. Elphic, * D. J. Lawrence, W. C. Feldman, B. L. Barraclough, S. Maurice, A. B. Binder, P. G. Lucey "Lunar Fe and Ti Abundances: Comparison of Lunar Prospector and Clementine Data" Science 4 September 1998:Vol. 281. no. 5382, pp. 1493 – 1496.

10. Sleep N. H., Zahle K. (1998) "Refugia From Asteroid Impacts on Early Mars and Early Earth" Jou. Geophys. Res. Vol 103, E12, 28529-28,544.

11. Win L. H. et al. "Long Term Sustainability of a High Energy, Low Diversity, Crustal Biome." Science Vol. 314, p479-482.

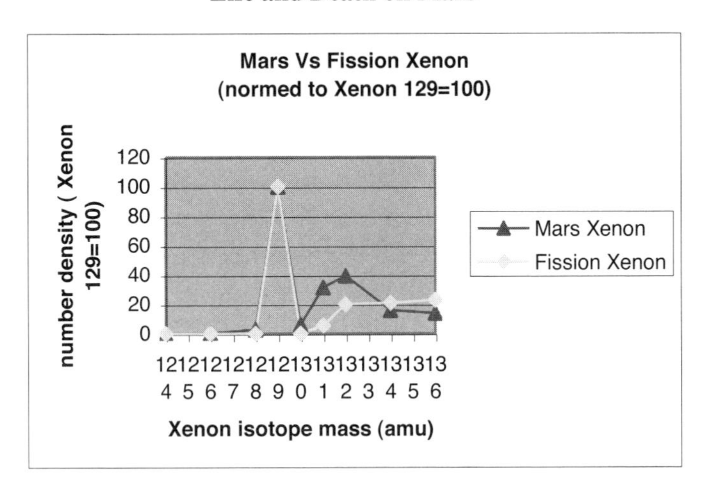

Figure 1. The Mass spectrum of stable isotopes in Mars atmosphere is shown in red, the corresponding mass spectrum for Earth's anthropgenic nuclear activity is is yellow. Both are normed to ^{129}Xe concentration =100.

The CI Carbonaceous Chondrites as the Missing Old Meteorites of Mars

Abstract

An array of data suggests the CI carbonaceous meteorites may be samples of early Mars regolith which is now preserved in the Southern highlands. Oxygen isotopes match well with aqueously altered materials from recognized meteorites as do anhydrous grains from the CI when compared to Martian lavas. Chromium isotopes also match well as do Nitrogen when compared to the CIs' recognized Martian contemporary, ALH84001. Igneous grains found in CIs fall on the same high Nickel and Ca-Mg mixing lines as recognized Mars meteorites. The aqueously altered clay and siderite materials match well with Mars soil analyses and similar minerals found in ALH84001. The absence of hypervelocity impact material, together with the inclusion of solar flare tracked grains, argues that the CI formed as a regolith in a velocity buffered environment that also supported abundant liquid water, conditions most easily achieved under a planetary atmosphere.

J. E. Brandenburg

Florida Space Institute-University of Central Florida
Kennedy Space Center

I. Introduction: the Mars Age Paradox

The absence of meteorites dating from Early Mars, except for ALH84001, is a continuing riddle. However, it now seems possible that the missing old MMs (Mars meteorites) may actually be already in human possession, but have not been recognized as Martian. The CI carbonaceous chondrites have been advanced as candidates for the missing old MMs (Brandenburg J. E., (1996)). These meteorites are, like ALH84001, 4.5Gyr in age. If they are from Mars they represent important probes of early Mars surface conditions. The possible identification of these meteorites as Martian has been done based on their detailed isotopic and chemical similarity to the known MMs. The CI are composed of clays and other water altered or soluble minerals and thus resemble not the recognized MMs, which are lavas, but rather strongly resemble the soil found at the Mars landing sites. This similarity of chemistry between CI material and the Viking soils was first noted by the investigators the Viking Mars lander soil analysis team (Toulmin et al. (1977)).

This possible identification of the CI as Martian was late in coming because the CI were, for a long time, thought to represent relics of the primordial solar nebula, and to be from a parent body in the asteroid belt. However, it was also recognized that their heavily aqueously altered state and absence of chondrules or any other evidence of hypervelocity impact argued for their origin in a unique planetary-like environment. Petrographic connections with the body of recognized MMs are certainly present: CI contain small grains which appear to be derived from a large lava melt similar to Martian lavas, and some MMs contain water altered materials similar to those found in the CI.

Despite the discovery of the one 4.5Gyr aged ALH84001 by Middlefehldt (1994) the rest of the MMs are young <1.2Gyr. The crustal dichotomy of Mars, with a young north and ancient south, would seem to require an approximately equal number of old to young meteorites in our collections. The process of ejection to Earth of ALH84001 was similar to the other MMs so ejection physics in South is the same as in North and no reason for a North-South isotropy in meteoritic bombardment has ever been suggested. Thus, some of our MMs are missing. Therefore, the absence of a large number of ancient meteorite material from the MMs collection is conspicuous and suggested that we may have unrecognized members of the MM collection in our possession. Do we then have MMs in our meteorite collections that, like ALH84001 initially, we do not recognize as Martian? The suggestion that the CIs might represent that group of missing MMs of sedimentary mineralogy, with origins in the southern portion of Mars was first made by Franchii et al. (1997). However, they noted what they believed to be a significant difference in oxygen isotopic composition between the CI and MMs, with CI materials being slightly higher than MM materials in $\Delta^{17}O$. In a full data set however, Martian aqueous materials were also found to have higher $\Delta^{17}O$ than MM lavas. Thus the perceived gap in oxygen isotope signatures between CI and MM materials has disappeared and with it the last substantive objections to the CI Mars connection.

II. Hypothesis

The CI-MM Hypothesis is that the CI carbonaceous chondrites are the missing old meteorites of Mars and are fragments of a sedimentary, water-altered, post-accretion "veneer" that formed on Mars during the Early Intense Bombardment period of Martian history, as was suggested by Anders and Owen (1977) to explain the noble gas isotopics found in the Mars atmsophere. That is, they are formed from the the "left-overs" of the formation of Mars from the the solar nebula, that rained down upon the newly formed surface of that planet and formed a veneer. This veneer, arriving late, never participated in the melt differentiation of Mars. This veneer is preserved in the Noachian South of Mars based on its old surface age but not in the North where resurfacing occurred late in Mars history. Fragments of the water altered veneer were ejected from Mars surface by recent impacts and conveyed to Earth and became the CIs (Brandenburg (1996)). The CI material never melt differentiated but was aqueously altered on Mars surface and sampled its environment.

If this CI-MM hypothesis is correct the CI and Mars should share the same isotopes and while never being melted they should be the same age as the primordial MM ALH84001. ALH84001 and CI are the same age ~ 4.5Gyr, if they are both from Mars then they sampled the same primordial environment and this should be reflected isotopically, chemically and morphologically. New data from ALH84001 supports this connection. Oxygen isotopic data for aqueously altered ALH84001 minerals (carbonates) is identical to CI aqueously altered silicates □^{17}O =0.8 found by Farquahar et al. (1998)and Baker et al. (1998). Nitrogen and Nobel gases closely matches CI isotopic composition for trapped gases in ALH84001 in 3g/cc mineral separates found by Murty and Mohpatra (1997).

In the remainder of this article, the full range of data isotopic, chemical, petro- morphological will be summarized as it relates to the MM-CI connection. The Implications of the MM-CI connection will be discussed, in particular the confirmation of organic matter on Mars.

III. Isotopic Data Relating to a Mars-CI Connection

Oxygen isotope data provides strongest connection between Mars and the CI. Oxygen is the third most abundant element in the Sun and is found in all rocks, thus oxygen isotope analysis, comparing the relative abundance of ^{17}O and ^{18}O normalized to the more abundant ^{16}O, is a valuable tool for differentiating meteorite samples, as shown in Figure 1. It is now apparent that the oxygen isotope signature of Mars is richer and more diverse than that of Earth, due to the fact that the hydro-sphere of Mars and its lithosphere are out isotopic equilibrium. The water of Mars is more enriched in the light isotope ^{17}O relative to its rock (Karlsson et al. 1992), this fact being attributed to the lack of plate tectonics on Mars. Tectonics is the major mechanism for isotopic equilibration on Earth because it brings subducted rock into contact with water under heat and pressure. Because of this dis-equilibrium of the anhydrous and hydrous minerals oxygen isotopic data in MM and CI materials must be compared in detail. It requires the comparison of anhydrous CI material with anhydrous MM material and likewise the comparison of aqueously altered materials between the two meteorite families.

For the CI the bulk of the material is aqueously altered, so the problem is to find aqueously deposited materials in the otherwise lava derived MMs. When this done comparison can be made between water deposited carbonates found in ALH84001 and other similar materials and the CI clay. When this is done, as shown in Figure 2, it is seen that corresponding CI and MM aqueously altered materials are indistinguishable.

In CI only small amounts of anhydrous materials are found, these consist of olivine and pyroxene grains imbedded in the clay matrix. The oxygen isotope data from the anhydrous mineral olivine and pyroxene grains from the CI (Watson L.A., Rubin A.E.,and McKeegan K.D. 1996) can be seen to be indistinguishable from a representative MM olivine (Figure 3).

Thus oxygen isotope data, the most fundamental differentiator between meteorites, has demonstrated a strong link exists between CI and MM s.

Light and Nobel Gas Isotopes

The measurement of another isotopic system D/H in water from both the MM and CI group is consistent with a common water reservoir for both groups. Measurement of the D/H ratio by ion microprobe in hydrous materials for three younger members of the MM group Chassigny, Shergotty, and Zagami, showed a wide range of values from □D = +4000 %o, which is the present Martian atmospheric value (Bjoraker et al. 1989) to +512%o, in water found in glass enclosed inclusions (Watson et al. 1994a). The range of D/H values was interested to reflect the intense fractionation of Mars hydrogen in time by the mechanism of water escape to space. The enclosed water of low □D was interpreted been the oldest water since it had been the most completely isolated from the atmosphere, where hydrogen is continually being fractionated. Measurements of D/H in water released by stepped heating, in the very ancient Martian meteorite, which formed contemporaneously with the CI Orgueil: ALH84001, which at 4.6 Gyr old (Jagoutz 1994) and was, showed also a 350C water release □D component of +33 %o and a 1000C release of + 700 %o (Watson et al 1994b), Watson et al. 1996, Boctor et al. 1998) showing lower D/H over all than other MMs (except for Chassigny, which gave almost terrestrial values using this method). This would appear reasonable, given ALH84001's greater age than other MMs. The values of □D measured in CI water samples released at 180C and above are + 170 to +235 %o inj Orgueil and + 180 to + 300 %o in Ivuna (Yang Epstein, 1982 and Boato, 1954).

The D/H ratio data for water in CI materials are thus consistent with exposure to water on Mars contemporaneously with ALH84001, when Mars water less fractionated.

Measurements of carbon isotopes in CI carbonates gave a □$^{13}C_{PDB}$ of +70.2 %o for Orguiel and +65.8 %o for Ivuna. These constitute the heaviest C measured in any meteoritic carbonate (Pillinger 1984). These can be compared with □$^{1□}C_{PDB}$ of +9 to +15 %o obtained from carbonates in two MMs : Nakhala and EETA 79001 (Wright et al. 1989) with ages of 1.3 Gyr or younger, and from ALH84001 +40.9 %o was obtained (Watson et al. 1994a), which has the heaviest carbon of any MM. This is consistent with a picture of an early Mars where fractionation of carbon was lessening in time and would again be consistent with CI carbonate formation on Mars contemporaneously with those found in ALH84001.

The CIs are remarkable in that they contain organic matter in greater abundance than any other meteoritic type (Nagy 1975). Because of this, a martian orgin for the CIs coupled with their great age, would mean evidence for primordial organo-synthesis on Mars and perhaps even primitive biology. However, this degree of organic matter content is is marked contrast to the apparent complete lack of detectable (1ppb) organic matter in the Martian soil at the Viking landing sites (Bieman and Lavoie, 1979) and the apparent hostility of the Martian Surface Environment to organic matter, due to solar UV and the presence of possible oxidizing agents in the Martian soil (Oyama and Berdahl, 1977). Thus the, the CI material would have to be material buried at depths of at least meters below the martian surface to be preserved over geologic time, out of contact with the

UV environment at the surface, but shallow enough to permit ejection.

Organic matter has been reported in some MMs and can be compared with that in CIs. The isotopic makeup of the solvent soluable organic carbon in the CI Ivuna is □$^{13}C_{PDB}$ = -18.0 %o (Pillinger 1984) which compares well with measured acid resisitant residue carbon in MMs □$^{13}C_{PDH}$ =-28.0 %o for Nahkala and □13CPDH =-33%o for EETA 79001 (Wright et al. 1989). Careful study of any identified primordial MM organics and comparison with CI organics could provide a good test of the hypothesis that the CI come from Mars, since under this hypothesis organic molecular species should be similar between the two.

Measurement of trapped noble gases and nitrogen in MM materials has provided important evidence for establishing the parent body of the MMs (McSween 1994, Bogard et al. 1984) Extensive analysis has revealed two components : a shock implanted "atmospheric " component matching the modern Mars atmospheric pattern of noble gases and nitrogen isotopic abundance and and a "solar" component thought to rpresent the dissolved gases in the Martian mantle (Swindle et al. 1986). These components appear mixed in several MM lithologies and thus Mars appears to present a rich spectrum of trapped gas isotopic signatures rather than a single one. In contrast, CI trapped gases follow a pattern of isotopic abundance belonging to a broad class called "planetary". At the heavier end of the isotope mass spectrum, Kr and Xe, the CI gases follow a nearly solar pattern called AVCC (Average Carboneacous Chondrites).

Amoung the Kr and Xe isotopes Swindle et al. have identified what appears to be a highly fractionated AVCC component in Shergotites (Swindle et al. 1986) and Drake et al. (1991) more recently has identified in MM trapped gases two mixing lines in the space of $^{86}Kr/^{132}Xe$ and $^{129}Xe/^{132}Xe$ abundance ratios. Both mixing lines have the data from Chassigny as an endpoint, indicating that it represents an important trapped gas component in these isotopes. The data from Chassigny, $^{86}Kr/^{132}Xe$= 1.14 and $^{129}Xe/\ ^{132}Xe$ =1.03. compares very favorably with data taken from Eugster et al. (1967) which gives $^{86}Kr/^{132}Xe$ = .98 and $^{129}Xe/^{132}Xe$ = 1.05 for Orgueil indicating that the CI and the MM Chassigny shared this same component. Orgueil magnetites have $^{129}Xe/^{132}Xe$ = 1.02 where this can have a contribution form the decay of ^{129}I. The Martian atmosphere has $^{129}Xe/^{132}Xe$ =2.5. These results seem consistent with the idea, suggested by Peppin (1992), that the early Martian atmosphere received a substantial contribution from CI like material.

Trapped nitrogen, released from the organic portion of the Orgueil meteorite, gives a □$^{15}N_{air}$ = +43 (Pillinger 1984) whereas nitrogen from EETA79001 glasses ranges from +90 < □$^{15}N_{air}$ < 225 (McSween 1985). Since Mars atmosphere is believed to have undergone fractionation in time and the CI are primordial in age, this slightly fractionated nitrogen in CIs is consistent with primordial Martian atmosphere.

This supported strongly by the discover of an end member nitrogen component in ALH84001, (Murty and Mohapatra 1997) which has of □$^{15}N_{air}$= +46. they also found a pattern of Xe isotopes that seems to be AVCC, they also found a Kr component in ALH84001 that has an isotopic signature of AVCC Kr (Figure 4).

In the lighter isotopes, gases trapped in Orgueil organic residues released above 900C, give average, $^{20}Ne/^{22}Ne$ = 8.5 and $^{38}Ar/^{36}Ar$= 0.19 (Frick and Monoit 1977) These can be compared with values obtained the Martian meteorite LEW 88516 released at 1600C which are : $^{20}Ne/^{22}Ne$ =0.92, $^{38}Ar/^{36}Ar$ =0.56, (Treiman et al. 1994). Corresponding values fro Sherggotite lithologies are in ranges 1< $^{20}Ne/^{22}Ne$ < 10 and .19 < $^{38}Ar/$ ^{36}Ar <1.0 (Bogard et al. 1984). Atmospheric values for Mars Mars Argon is $^{38}Ar/^{36}Ar$=0.18 Therefore, in a limited analysis presented here, trapped gas data would appear to be consistent with a common origin for both MM and CI noble gases and nitrogen, as well as oxygen.

Therefore the light, stable isotopic data for both water and lithologies are consistent with MM and CI materials having a common origin.

Tungsten and Chromium

Recent isotopic studies of MM meteorite have revealed a chromium and tungsten signature for MMs. For Chromium the comparison between recognized Mars meteorites is informed again by the fact that the MMs represent rock from a chemically differentiated lava melt whereas the CI represent a aqueously processed mixture of materials from a planetary nebula, that simply arrived on Mars too late to be part of the melt. The CI have long been recognized as being composed of a range of materials whose Cr isotopes are individually quite different from those characteristic of Mars or Earth but whose range subsumes those values. A major component of the CI chromium does in fact lie on the Mars value (Endress et al. (1996)). (Figure 5)

For tungsten the signature relates the time of molten core differentiation to the time of decay of short lived isotopes (Foley C.N. et al. (2003))The agreement in other isotopes with a gap between the range of tungsten isotopic values and those for the CI is consistent with the CI being a late accretion veneer on Mars that did not participate in melt differentiation. Thus, unlike the MMs who formed from a large lava melt and chemically differentiated while the short lived ^{182}Hf was still "live", the CI remained in their undifferentiated state and preserved primordial ratios of isotopes. (Figure 6)

Space Exposure Ages

Like all meteorites, the CI are believed to have been buried on the surface of some larger parent body, where they were protected from cosmic rays, when they are ejected by some other impact they are then exposed to cosmic rays which create "spallation" product isotopes. The isotopic composition of light noble gases such as helium and neon, products of cosmic ray spallation, is an important indicator of exposure time to interplanetary space. Estimates of the space exposure times of MM and CI meteorites are shown in Figure 7 and can thus be compared. They are appear to be comparable.

Therefore, it is apparent that the CI and MM meteorites came from the same distinct reservoir of isotopes. However, the same can be said for both lunar and terrestrial samples, so that the question then becomes, did the MM and CIs sample the same geo-chemical conditions? This can be answered by comparing the minerals found in them.

IV. Mineralogical and Chemical Comparisons between CI and MM groups.

Despite the fact that the MMs are overwhelmingly igneous rock and the CI are water formed clay, comparisons can be made between the groups and between the approximately known Martian soils from the landing sites. Igneous rock also appears in the CI in the form of small pyroxene and olivine grains.

Olivine grains are found in carbonaceous chondrites generally, being a component of the dust of the early planetary nebula. However, the olivines found in the CI are unique for those found in carbonaceous chondrites and these unique properties suggest their origin from a planetary sized parent body. The olivine grains found in the CI are unique for those found in carbonaceous chondrites in that the iron rich grains follow a mixing line in Calcium and Magnesium (Kerridge, J. F. and MacDougall J.D. 1976). This is considered remarkable since calcium rich and magnesium rich minerals have very different melting and condensation points and thus the grain cannot be formed directly from a solar nebula. Instead the grains appear to have been formed as droplets splattered out of a lava melt on a large body. The olivines appear to fall on the same mixing line as Chassigny, the Martian olivine. (see Figure 8)

In addition, the iron rich olivine grains from CI are rich in nickel. This is considered important because nickel enrichment of olivines is considered to occur only on large bodies, due to the expulsion of nickel from the iron rich molten cores. It can be seen that the olivines from CI fall on the same mixing line as Martian olivines (see Figure 9)

CI can be said to resemble Martian landing site soils in that they are also water formed clays. The similarity between CI and Martian soil composition was striking and was noted at the first landings, to the extent that a Martian origin for CI material was then actively considered (Toulmin et al. 1979). However, since at the time no meteorite was known to have

Mars as a parent body, and the CI material was so fragile, this possibility was immediately discounted (Klause Kyle Private communication).

One aqueously formed mineral found in CI and MMs is quite distinct in both groups and is not found in other meteorite types. Ferrous magnesite, also called brunnerite, is found in only two meteoritic types in any abundance: MMs and CIs. The composition of CI brunnerite closely matches that found that ALH84001 .(Figure 10)

Therefore, it appears that MM and CI meteorites formed not only from the same pot of isotopes but were formed under the same chemical conditions. The simplest hypothesis then becomes that they both originated on the same planetary body.

V. Petro-Morphology

The CI, like almost all meteorites, were once buried away from cosmic ray bombardment below the surface of a parent body somewhere in the solar system. This parent body was impacted and then released the CI as secondary fragments that made their way to Earth. The morphology of the CI offers constraints on the nature of the parent body.

The CI are described as composed of clay that was brecciated by low velocity impact and with the clasts then cemented back together by water soluble salts, chiefly carbonates, with later sulfate formation. The individual clay clasts have a pronounced lamellar character (Figure 11), suggesting that they formed in a strong gravity field. There is no evidence for hypervelocity impact. In this brecciated mix isolated olivine grains are found, some with solar flare tracks, indicating that they were once in solar space. The olivines contain both a low iron forsteritic branch and a high iron, high nickel, branch. The latter population strongly differentiates the olivine grains from those found in CM chondrites. This is a further differentiation from CM chondrites where there are abundant chondrites and much evidence for hypervelocity impact. The olivine grains in CI are unique also in that they are often embayed, with embayments concentrated in iron rich portions of the grains.

The CI are also, as a group remarkably uniform in composition and morphology, indicating a large parent body, capable of enforcing large scale uniformity on its surface.

Based on the presence of tracked olivine grains, the CI clearly formed as regolith on some parent body. However, based on the absence of hypervelocity impact they were formed in velocity buffered environment. The simplest way for this to occur is that the CI formed under an atmosphere, like a rock on Earth. The parent body of the CI was thus massive enough to form and hold an atmosphere and support the abundance of water they were exposed to. The gravity of such a body would also easily explain the lamellar character of the clasts.

Embayments are found on some CI olivine grains (Kerridge, J. F. and MacDougall J.D. 1976), coinciding with the lower temperature melting iron rich portions. The presence of such embayments can be explained as being caused by ablation as they fell through an atmosphere (Figure 12). Accordingly, the simplest model for formation of the CI parent minerals was they formed on a massive parent body. This parent body had large lava melts and these served as sources for olivine grains when impacted, accounting for the calcium-iron mixing line present in the iron rich olivine grains.

Therefore, the morphology of the CI and their complex history of several periods of inundation in water, followed by desiccation, brecciation by mild impact, then re- immersion in water, is most simply explained by their formation in a planetary environment similar to early Mars, rather than some minor body in the solar system.

V. Discussion: the CI connection to Mars

Isotopically, and chemically, the CI appear to be Martian in origin. CI uniquely have no chondrules or evidence of hypervelocity impact yet, as evidenced by solar tracked grains, were a regolith material. This is consistent only with formation under atmosphere which stops chondrules but not grains and provides velocity buffering. Additionally, CI grains are uniquely embayed-zoned suggesting ablation in atmosphere. The simplest explanation for this array of features is that the CI formed under atmosphere on Mars rather than some minor body in the asteroid belt.

If the CI-MM connection is real then Martian age paradox is solved. The CI are the missing old meteorites of Mars and this discovery also practically doubles our amount of Martian material. It also tells us the south of Mars is very ancient and also heavily water altered and that impacts have supplied us with materials from both hemispheres uniformly. It also confirms the model of cratering statistics proposed by Nadine Barlow (1988), who proposed that the surface ages of Mars were largely bi-modal, with very ancient terrains in the south of the dichotomy and young terrains to the north. Mars southern hemisphere nearly devoid of aqueous and geological activity since Early Intense Bombardment Era.

Models of Mars accretion from CI like material are supported as is CI veneer on Mars (Pepin 1992, Dreibus and Wanke, 1985). The hypothesis that Mars is parent body of CI carbonaceous chondrites, is a revival of an earlier hypothesis of a planetary seabed origin of CI material made by Urey (1968). The high levels of organic a material found in the CI argues for an early Martian surface that was warm, wet, and rich in organics.

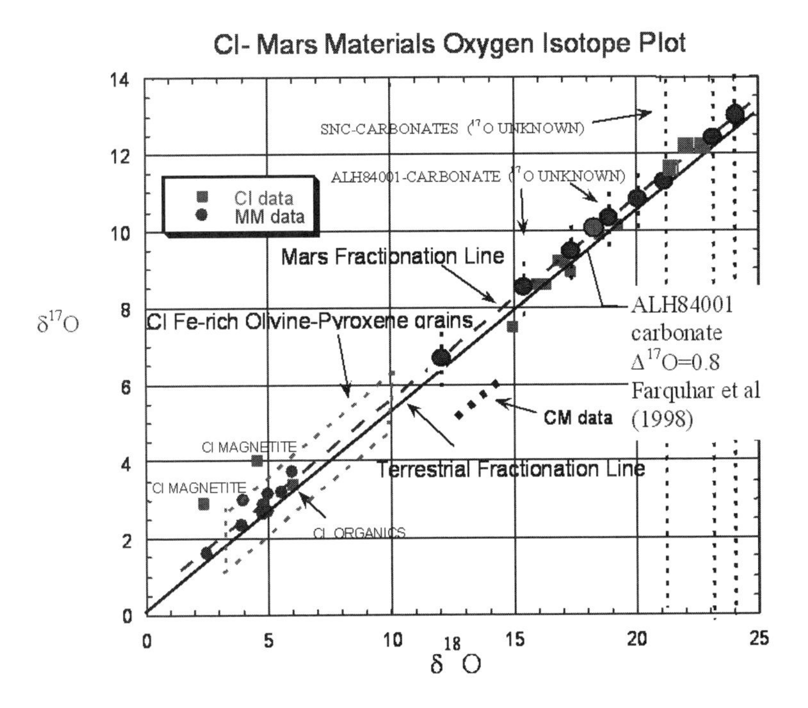

Figure 1. Oxygen isotopic data for meteorites. MM meteorites are Nakhla and Lafayette. Solid line is the terrestrial fractionation line where terrestrial and lunar materials are found, dashed line is that for Martian materials

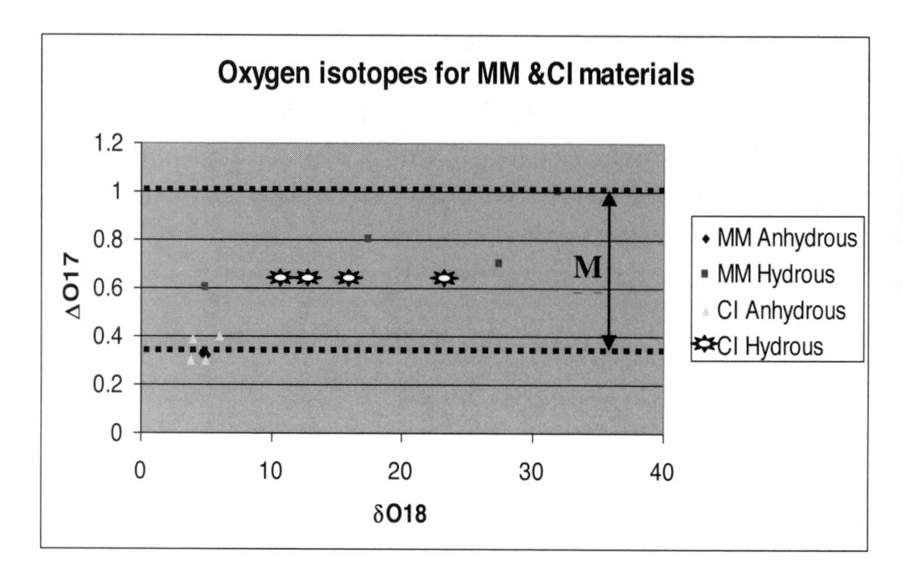

Figure 2. A comparison of oxygen isotopic data for aqueous and anhydrous minerals from both MMs and CIs. They are compared using the □^{17}O verus □$^{1□}$O which measures difference in parts per mil from the terrestrial fractionation line and parts per mil difference from terrestrial ocean water. Dashed lines represent upper and lower bounds of Mars meteorite materials.

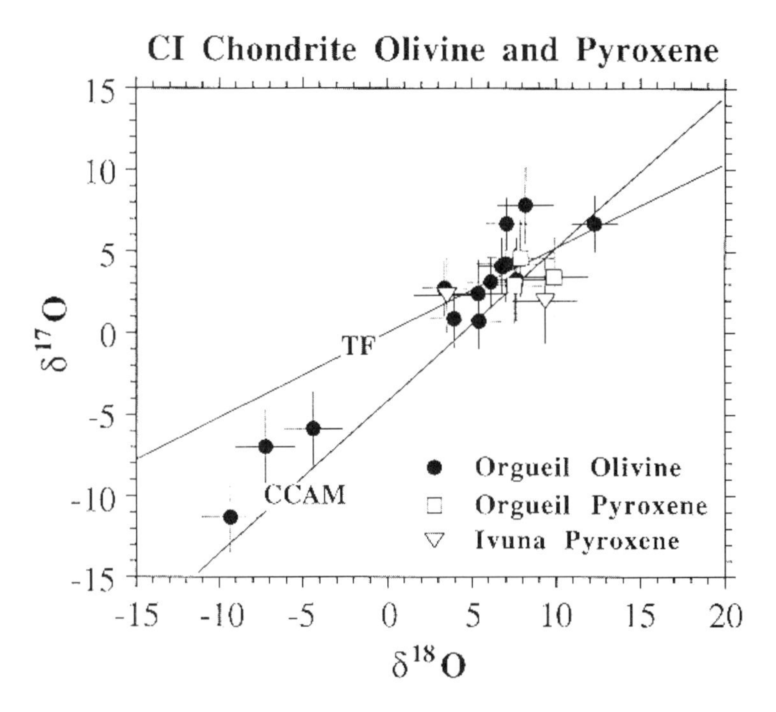

Figure 3. A graph of oxygen isotopes for CI olivine and pyroxene grains compared to a representative MM olivine form the MM Chassigny.

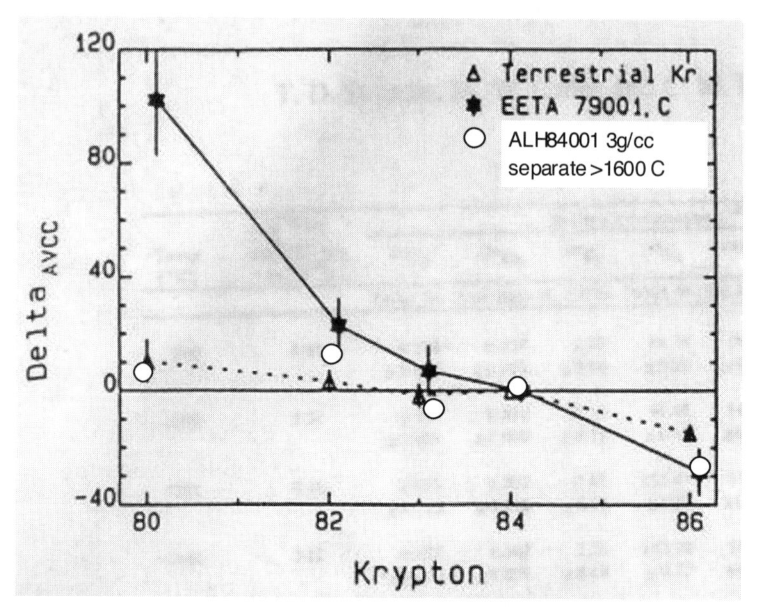

Figure 4. CI versus Martian Kr (graph adapted from Swindle, T. D., Caffee, M. W., and Hohenberg, C. M., (1986))

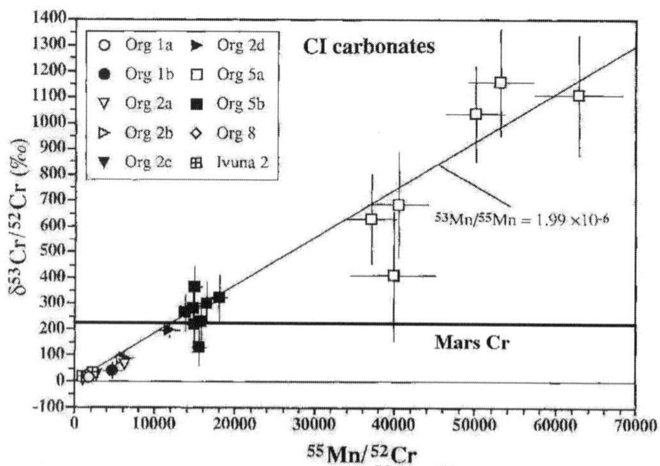

Figure 5. A comparison of $^{53}Cr/^{52}Cr$ for CI materials also showing the recently identified MM signature value. Graph adapted from Magnus et al. (1996)

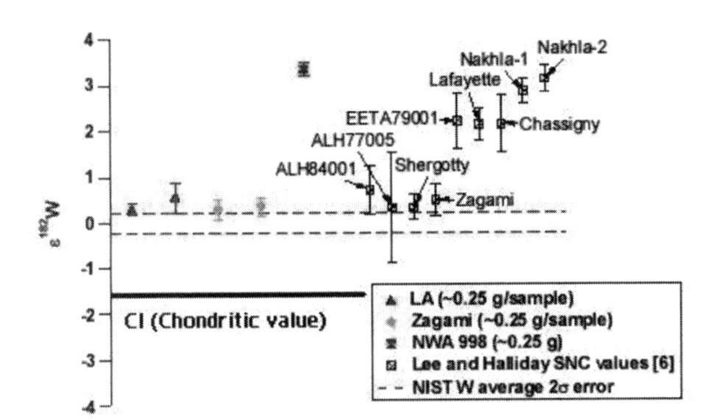

Figure 6. A comparison of Mars and Terrestrial Tungsten isotopes with those of CI.

Graph adapted from Foley et al (2003)

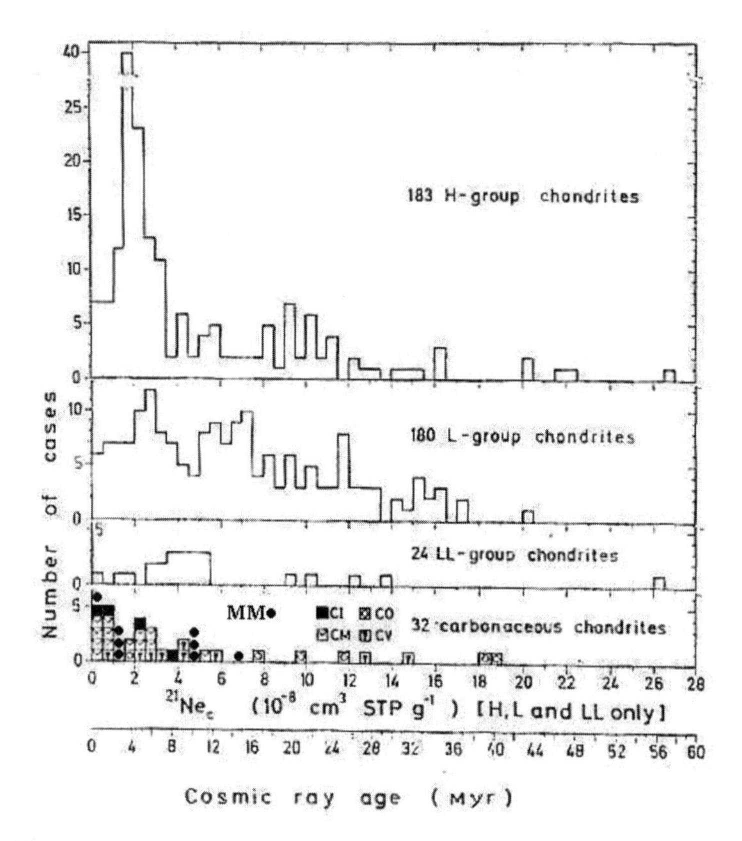

Figure 7. a comparison of cosmic ray exposure ages for CI and presently recognized MMs as well as other meteorite types. Graph adapted from Asteroids, ed. T. Gerhels The University of Arizona Press, Tucson & London p. 544.

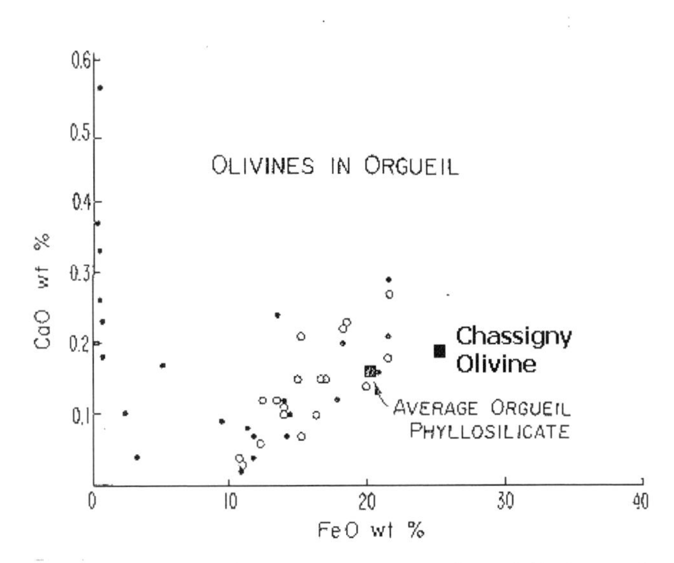

Figure 8. CI Olivines versus Martian Olivines Ca-Fe mixing Line (adapted from Kerridge, J. F. and MacDougall J.D. 1976)

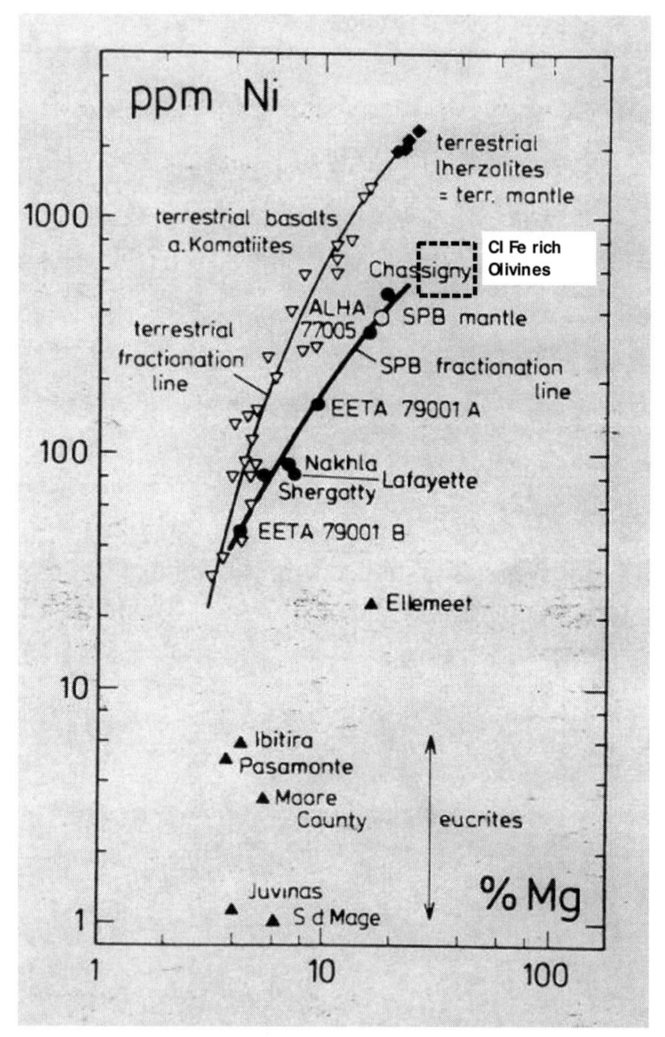

Figure 9. Nickel content of iron rich CI Olivines versus Martian Olivines. (adapted from Kerridge, J. F. and MacDougall J.D. 1976)

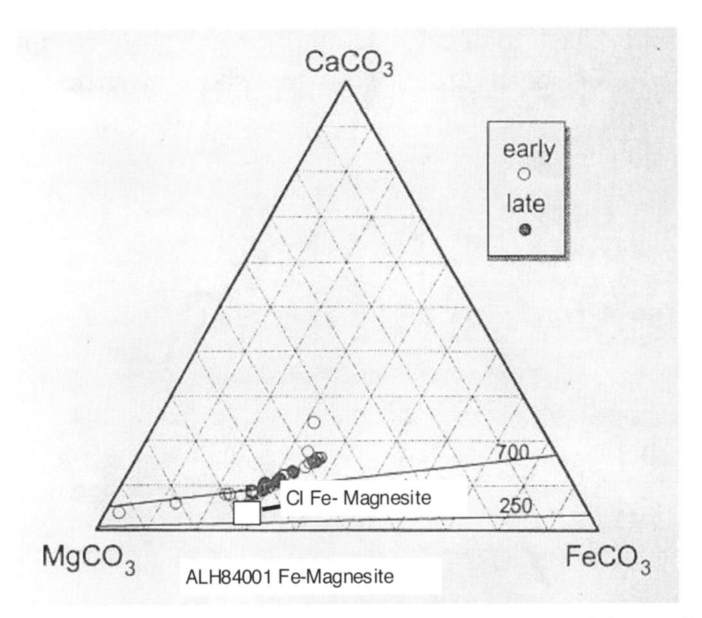

Figure 10. A comparison of the composition of ferroan magnesite found in ALH84001 and that found its CIs (adapted from Wright I.P., M.M. Grady, and C.T. Pillinger. 1992).

Figure 11 Clast in the CI Chondrite Orgueil, composed mostly of ill defined hydrated silicates, showing pronounced lamellar texture. Width of field of view is 1 mm. (from Kerridge and Bunch)

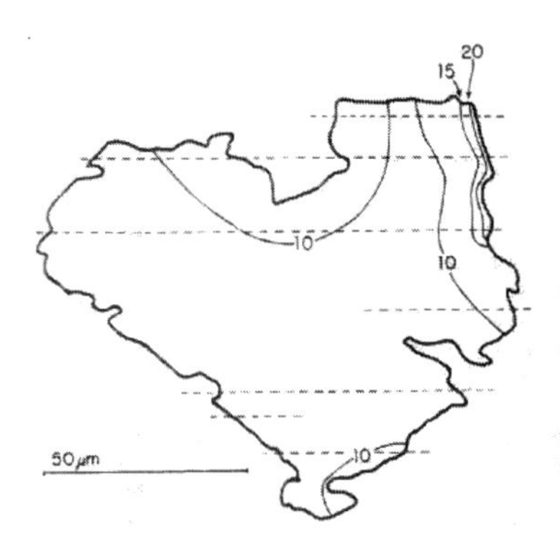

Figure 12 Outline of a zoned olivine grain from Orgueil showing contours of iron (and calcium) enrichment. Valuse give FeO content in wt. % Dashed lines show locations of microprobe traverses. (from Kerridge and MacDougall 1976)

References

Anders, E., and Owen T. (1977) " Mars and Earth: Origin and Abundance of Volatiles" Science 198:453-465.

Baker L. et al. (1998) "Measurement of Oxygen isotopes in Water from CI and CM Chondrites." LPI Conf. 1998.

Barlow N. G.(1988) "Crater Size-Frequency Distribution and a Revised Martian Chronology" Icarus 75,pp285-305.

Boctor N. Z. et al. (1998) "Petrology and hydrogen and Sulfur Isotope Studies of mineral Phases in Martian Meteorite ALH84001 Lunar and Planetary Science Conference 1998.

Brandenburg J. E., (1996), Mars as the Parent Body of the CI Carbonaceous Chondrites, Geophys. Res. Lett., 23,9, p.961-964

Clayton, Robert N.,and Mayeda, Toshiko K. (1983), Oxygen isotopes in eucrites, shegottites, nakhlites, and chassignites, Earth and Planetary Science Letters, 62, pp1-6.

Dreibus, G., and Wanke, H. (1985) Mars, a volatile-rich planet. Meteoritics, 20 pp.367-381

Endress, Magnus, Zinner, Erenst, and Bischoff, Adolf, (1996) " Early Aqueous activity on Primitive Meteorite Parent Bodies, Nature 379, p701.

Farquahar J., Theimens M.H. and Jackson T. (1998) "D17O Measurements of Carbonate from ALH84001: Implications for

Oxygen Cycling between the Atmosphere -Hydrosphere and Pedosphere of Mars" LPI Conf. 1998.

Floran R..J. et. al. (1978) The Chasssigny meteorite: A cumulate dunite with hydrous amphibole-bearing melt inclusions " Geochim Cosmochim Acta 42:pp.1213-1229

Fredrickson Kurt and Kerridge J.F. (1988) Carbonates and Sulfates in CI Chondrites: Formation by Aqueous Activity on the Parent body, Meteoritics, 23,pp35-44

Foley C.N., Wadhwa M.,and Janney P.E.,(2003) "Tungsten Isotopic Compositions of the SNC Meteorites: Further Implications for Early Differentiation History of Mars." Sixth International Conference on Mars (absract)

Karlsson, Haraldur R.,Clayton R.N., Gibson, Everett K. Jr., and Mayeda T.K.(1992), Water in SNC meteorites: Evidence for a Martian hydrosphere, Science, 255, pp.1409-1411.

Kerridge, J. F. Bunch, T.E., (1979) Aqueous alteration on asteroids: Evidence from carbonaceous meteorites. In Asteroids, ed. T. Gerhels The University of Arizona Press, Tucson & London pp. 610-611.

Kerridge, J. F. and MacDougall J.D. (1976) Mafic silicates in the Orgueil Carbonaceous Meteorite, Earth and Planetary Science Letters, 29,pp341-348.

MacDougall J.D.and Lugmair G.W.(1984), Early solar system aqueous activity: Sr isotope evidence from the Orgueil CI meteorite, Nature,307,pp.249-251,.

McSween H., Y. (1977) On the nature and origin of isolated olivine grains in carbaceous chondrites, Geochim. Cosmochim. Act., 58,23, pp5341-5347.

Murty S.V.S. and Mohapatra R.K. (1997) Nitrogen and heavy Isotopes in ALH84001, Geochim.etCosmochim. Acta 61,24,5417-5428

Mittlefehldt, D.W.(1994) ALH84001,a culminate orthopryoxenite member of the SNC meteorite group. Meteoritics, 29, 214-221.

Nagy, B. (1975) Carbonaceous Chondrites, Elsevier Amsterdam

Pepin, Robert (1992) The Origin of Nobel Gases in Terrestrial Planets, Annual Review of Earth and Planetary Science, 20, pp389-430.

Pillinger C.T. (1984) “Light Element Stable Isotopes in Meteorites- From Grams to Picograms” GeoChim. CosmoChim. Acta 48 pp2739-2766

Rowe Marvin W., Clayton R.N., and Mayeda T.K.(1994), Oxygen isotopes in separated components of CI and CM meteorites, Geochim. Cosmochim. Act., 58,23, pp5341-5347.

Urey Harold C., (1968) Origin of some meteorites from the Moon, Naturwiss Enshaften,55, pp 177-181.

Swindle, T. D., Caffee, M. W., and Hohenberg, C. M., (1986) "Xenon and other noble gases in shergottites" Geochimica et Cosmochimica Acta, 50, pp 1001-1015.

Tanaka K.L. (1986) "The Stratigraphy of Mars" Proc. Lunar Planet. Sci. Conf. 17, Jour. Geophys. Res. Suppl. 91:E139-E158.

Toulmin, Priestly III, A.K. Baird, B. Clark, K. Keil, H.J. Rose, Jr. R. P.

Christian, P.H. Evans, and W.C. Kelliher (1977), Goechemical and minerological interpetation of the Viking inorganic chemical results, Jour. Geo. Res. 82,28, pp4625-4633.

Treiman A.H., McKay G.A., Bogard D.D., Mittlefehldt D.W., Wang M.S., Keller L., Lipshcultz M.E., Lindstrom M.M., Garrison D.(1994) Comparison of the LEW88516 and ALHA77005 martian meteorites: Similar but distinct, Meteoritics,29,pp.581-592.

Watson L.A., Rubin A.E., and McKeegan K.D. (1996) Oxygen isotopic compositions of Olivine and Pyroxene form CI meteorites. LPSC XXVII, p745-746 (abstract)

Watson L.L., Hutcheon I. D., Epstein S., and Stolper E. M., (1994a) Water on Mars: Clues from deuterium /hydrogen and water contents of hydrous phases in SNC meteorites. Science, 265, pp. 86-89.

Watson L.L., Epstein S., and Stolper, E. M. (1994b) D/H of water released by stepped heating of Shergotty, Zagami, Chassigny, and Nakhla. Meteoritics, 29, p. 547. (Abstract)

Wright I.P., M.M. Grady, and C.T. Pillinger(1992), Chassigny and the nakhlites: carbon bearing components and their relationship to martian envirnmental conditions, Geochimica et Cosmochimica Acta, 56, pp 817-826.

Valley J.W. et al. (1997) "low temperature Carbonate Concretions in the Martian Meteorite ALH84001: Evidence from Stable isotopes and Minerology " Science,275,pp1633-1638

Yang J. and Epstein S. (1982) On the origin and composition of hydrogen and carbon in meteorites, Meteoritics, 17,p301

WEATHER WARFARE
The Military's Plan to Draft Mother Nature
by Jerry E. Smith

Weather modification in the form of cloud seeding to increase snow packs in the Sierras or suppress hail over Kansas is now an everyday affair. Underground nuclear tests in Nevada have set off earthquakes. A Russian company has been offering to sell typhoons (hurricanes) on demand since the 1990s. Scientists have been searching for ways to move hurricanes for over fifty years. In the same amount of time we went from the Wright Brothers to Neil Armstrong. Hundreds of environmental and weather modifying technologies have been patented in the United States alone – and hundreds more are being developed in civilian, academic, military and quasi-military laboratories around the world *at this moment!* Numerous ongoing military programs do inject aerosols at high altitude for communications and surveillance operations.

304 Pages. 6x9 Paperback. Illustrated. Bib. $18.95. Code: WWAR

HAARP
The Ultimate Weapon of the Conspiracy
by Jerry Smith

The HAARP project in Alaska is one of the most controversial projects ever undertaken by the U.S. Government. Jerry Smith gives us the history of the HAARP project and explains how works, in technically correct yet easy to understand language. At at worst, HAARP could be the most dangerous device ever created, a futuristic technology that is everything from super-beam weapon to world-wide mind control device. Topics include Over-the-Horizon Radar and HAARP, Mind Control, ELF and HAARP, The Telsa Connection, The Russian Woodpecker, GWEN & HAARP, Earth Penetrating Tomography, Weather Modification, Secret Science of the Conspiracy, more. Includes the complete 1987 Eastlund patent for his pulsed super-weapon that he claims was stolen by the HAARP Project.

256 pages. 6x9 Paperback. Illustrated. Bib. $14.95. Code: HARP

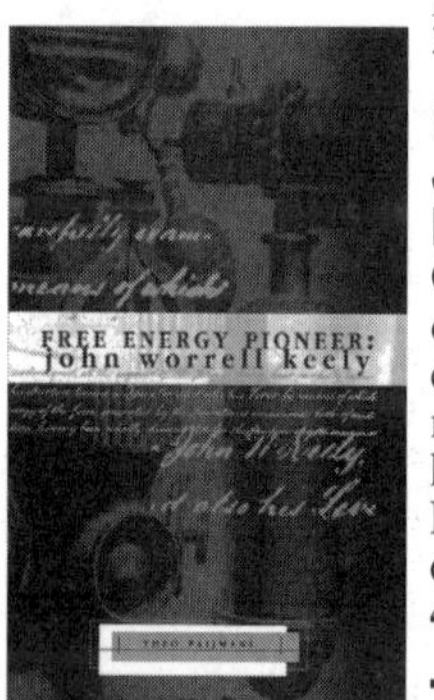

FREE ENERGY PIONEER
John Worrell Keely
by Theo Paijmans, foreword by John A. Keel

Over a century ago, a man in Philadelphia made the most important discovery of all time: a mysterious source of free, unlimited energy. He experimented with the substance for years, building a staggering 2,000 machines and devices that ran on his esoteric force. His eccentric vision led him to experiment with anti-gravity and the disintegration of solid matter. Lots of illustrations; Keely's mysterious source of unlimited, everlasting energy; Keely's many inventions; more.

416 pages. 6x9 Paperback. Illustrated. Bib. Index. $19.95. code: FEP

THE FANTASTIC INVENTIONS OF NIKOLA TESLA

by Nikola Tesla with David Hatcher Childress

This book is a readable compendium of patents, diagrams, photos and explanations of the many incredible inventions of the originator of the modern era of electrification. In Tesla's own words are such topics as wireless transmission of power, death rays, and radio-controlled airships. In addition, rare material on a secret city built at a remote jungle site in South America by one of Tesla's students, Guglielmo Marconi. Marconi's secret group claims to have built flying saucers in the 1940s and to have gone to Mars in the early 1950s! Incredible photos of these Tesla craft are included. •His plan to transmit free electricity into the atmosphere. •How electrical devices would work using only small antennas. •Why unlimited power could be utilized anywhere on earth. •How radio and radar technology can be used as death-ray weapons in Star Wars.

342 PAGES. 6x9 PAPERBACK. ILLUSTRATED. $16.95. CODE: FINT

SECRETS OF THE HOLY LANCE
The Spear of Destiny in History & Legend
by Jerry E. Smith

Secrets of the Holy Lance traces the Spear from its possession by Constantine, Rome's first Christian Caesar, to Charlemagne's claim that with it he ruled the Holy Roman Empire by Divine Right, and on through two thousand years of kings and emperors, until it came within Hitler's grasp—and beyond! Did it rest for a while in Antarctic ice? Is it now hidden in Europe, awaiting the next person to claim its awesome power? Neither debunking nor worshiping, *Secrets of the Holy Lance* seeks to pierce the veil of myth and mystery around the Spear. Mere belief that it was infused with magic by virtue of its shedding the Savior's blood has made men kings. But what if it's more? What are "the powers it serves"?

312 PAGES. 6X9 PAPERBACK. ILLUSTRATED. BIBLIOGRAPHY. $16.95. CODE: SOHL

PRODIGAL GENIUS
The Life of Nikola Tesla
by John J. O'Neill

This special edition of O'Neill's book has many rare photographs of Tesla and his most advanced inventions. Tesla's eccentric personality gives his life story a strange romantic quality. He made his first million before he was forty, yet gave up his royalties in a gesture of friendship, and died almost in poverty. Tesla could see an invention in 3-D, from every angle, within his mind, before it was built; how he refused to accept the Nobel Prize; his friendships with Mark Twain, George Westinghouse and competition with Thomas Edison. Tesla is revealed as a figure of genius whose influence on the world reaches into the far future. Deluxe, illustrated edition.

408 pages. 6x9 Paperback. Illustrated. Bibliography. $18.95. Code: PRG

TAPPING THE ZERO POINT ENERGY
Free Energy & Anti-Gravity in Today's Physics
by Moray B. King

King explains how free energy and anti-gravity are possible. The theories of the zero point energy maintain there are tremendous fluctuations of electrical field energy imbedded within the fabric of space. This book tells how, in the 1930s, inventor T. Henry Moray could produce a fifty kilowatt "free energy" machine; how an electrified plasma vortex creates anti-gravity; how the Pons/Fleischmann "cold fusion" experiment could produce tremendous heat without fusion; and how certain experiments might produce a gravitational anomaly.

180 PAGES. 5X8 PAPERBACK. ILLUSTRATED. $12.95. CODE: TAP

QUEST FOR ZERO-POINT ENERGY
Engineering Principles for "Free Energy"
by Moray B. King

King expands, with diagrams, on how free energy and anti-gravity are possible. The theories of zero point energy maintain there are tremendous fluctuations of electrical field energy embedded within the fabric of space. King explains the following topics: TFundamentals of a Zero-Point Energy Technology; Vacuum Energy Vortices; The Super Tube; Charge Clusters: The Basis of Zero-Point Energy Inventions; Vortex Filaments, Torsion Fields and the Zero-Point Energy; Transforming the Planet with a Zero-Point Energy Experiment; Dual Vortex Forms: The Key to a Large Zero-Point Energy Coherence. Packed with diagrams, patents and photos.

224 PAGES. 6X9 PAPERBACK. ILLUSTRATED. $14.95. CODE: QZPE

ATLANTIS & THE POWER SYSTEM OF THE GODS

by David Hatcher Childress and Bill Clendenon

Childress' fascinating analysis of Nikola Tesla's broadcast system in light of Edgar Cayce's "Terrible Crystal" and the obelisks of ancient Egypt and Ethiopia. Includes: Atlantis and its crystal power towers that broadcast energy; how these incredible power stations may still exist today; inventor Nikola Tesla's nearly identical system of power transmission; Mercury Proton Gyros and mercury vortex propulsion; more. Richly illustrated, and packed with evidence that Atlantis not only existed—it had a world-wide energy system more sophisticated than ours today.

246 PAGES. 6x9 PAPERBACK. ILLUSTRATED. $15.95. CODE: APSG

TECHNOLOGY OF THE GODS

The Incredible Sciences of the Ancients

by David Hatcher Childress

Childress looks at the technology that was allegedly used in Atlantis and the theory that the Great Pyramid of Egypt was originally a gigantic power station. He examines tales of ancient flight and the technology that it involved; how the ancients used electricity; megalithic building techniques; the use of crystal lenses and the fire from the gods; evidence of various high tech weapons in the past, including atomic weapons; ancient metallurgy and heavy machinery; the role of modern inventors such as Nikola Tesla in bringing ancient technology back into modern use; impossible artifacts; and more.

356 PAGES. 6x9 PAPERBACK. ILLUSTRATED. BIBLIOGRAPHY. $16.95. CODE: TGOD

VIMANA AIRCRAFT OF ANCIENT INDIA & ATLANTIS

by David Hatcher Childress, introduction by Ivan T. Sanderson

In this incredible volume on ancient India, authentic Indian texts such as the *Ramayana* and the *Mahabharata* are used to prove that ancient aircraft were in use more than four thousand years ago. Included in this book is the entire Fourth Century BC manuscript *Vimaanika Shastra* by the ancient author Maharishi Bharadwaaja. Also included are chapters on Atlantean technology, the incredible Rama Empire of India and the devastating wars that destroyed it.

334 PAGES. 6x9 PAPERBACK. ILLUSTRATED. $15.95. CODE: VAA

GRAVITATIONAL MANIPULATION OF DOMED CRAFT

UFO Propulsion Dynamics

by Paul E. Potter

Potter's precise and lavish illustrations allow the reader to enter directly into the realm of the advanced technological engineer and to understand, quite straightforwardly, the aliens' methods of energy manipulation: their methods of electrical power generation; how they purposely designed their craft to employ the kinds of energy dynamics that are exclusive to space (discoverable in our astrophysics) in order that their craft may generate both attractive and repulsive gravitational forces; their control over the mass-density matrix surrounding their craft enabling them to alter their physical dimensions and even manufacture their own frame of reference in respect to time. Includes a 16-page color insert.

624 pages. 7x10 Paperback. Illustrated. References. $24.00. Code: GMDC

THE TESLA PAPERS

Nikola Tesla on Free Energy & Wireless Transmission of Power

by Nikola Tesla, edited by David Hatcher Childress

David Hatcher Childress takes us into the incredible world of Nikola Tesla and his amazing inventions. Tesla's fantastic vision of the future, including wireless power, anti-gravity, free energy and highly advanced solar power. Also included are some of the papers, patents and material collected on Tesla at the Colorado Springs Tesla Symposiums, including papers on: •The Secret History of Wireless Transmission •Tesla and the Magnifying Transmitter •Design and Construction of a Half-Wave Tesla Coil •Electrostatics: A Key to Free Energy •Progress in Zero-Point Energy Research •Electromagnetic Energy from Antennas to Atoms •Tesla's Particle Beam Technology •Fundamental Excitatory Modes of the Earth-Ionosphere Cavity

325 PAGES. 8x10 PAPERBACK. ILLUSTRATED. $16.95. CODE: TTP

UFOS AND ANTI-GRAVITY

Piece For A Jig-Saw

by Leonard G. Cramp

Leonard G. Cramp's 1966 classic book on flying saucer propulsion and suppressed technology is a highly technical look at the UFO phenomena by a trained scientist. Cramp first introduces the idea of 'anti-gravity' and introduces us to the various theories of gravitation. He then examines the technology necessary to build a flying saucer and examines in great detail the technical aspects of such a craft. Cramp's book is a wealth of material and diagrams on flying saucers, anti-gravity, suppressed technology, G-fields and UFOs. Chapters include Crossroads of Aerodymanics, Aerodynamic Saucers, Limitations of Rocketry, Gravitation and the Ether, Gravitational Spaceships, G-Field Lift Effects, The Bi-Field Theory, VTOL and Hovercraft, Analysis of UFO photos, more.

388 PAGES. 6x9 PAPERBACK. ILLUSTRATED. $16.95. CODE: UAG

THE COSMIC MATRIX

Piece for a Jig-Saw, Part Two

by Leonard G. Cramp

Cramp examines anti-gravity effects and theorizes that this super-science used by the craft—described in detail in the book—can lift mankind into a new level of technology, transportation and understanding of the universe. The book takes a close look at gravity control, time travel, and the interlocking web of energy between all planets in our solar system with Leonard's unique technical diagrams. A fantastic voyage into the present and future!

364 PAGES. 6x9 PAPERBACK. ILLUSTRATED. BIBLIOGRAPHY. $16.00. CODE: CMX

THE A.T. FACTOR

A Scientists Encounter with UFOs

by Leonard Cramp

British aerospace engineer Cramp began much of the scientific anti-gravity and UFO propulsion analysis back in 1955 with his landmark book *Space, Gravity & the Flying Saucer* (out-of-print and rare). In this final book, Cramp brings to a close his detailed and controversial study of UFOs and Anti-Gravity.

324 PAGES. 6x9 PAPERBACK. ILLUSTRATED. BIBLIOGRAPHY. INDEX. $16.95. CODE: ATF

MAPS OF THE ANCIENT SEA KINGS
Evidence of Advanced Civilization in the Ice Age
by Charles H. Hapgood

Charles Hapgood has found the evidence in the Piri Reis Map that shows Antarctica, the Hadji Ahmed map, the Oronteus Finaeus and other amazing maps. Hapgood concluded that these maps were made from more ancient maps from the various ancient archives around the world, now lost. Not only were these unknown people more advanced in mapmaking than any people prior to the 18th century, it appears they mapped all the continents. The Americas were mapped thousands of years before Columbus. Antarctica was mapped when its coasts were free of ice!

316 pages. 7x10 paperback. Illustrated. Bibliography & Index. $19.95. code: MASK

PATH OF THE POLE
Cataclysmic Pole Shift Geology
by Charles H. Hapgood

Maps of the Ancient Sea Kings author Hapgood's classic book *Path of the Pole* is back in print! Hapgood researched Antarctica, ancient maps and the geological record to conclude that the Earth's crust has slipped on the inner core many times in the past, changing the position of the pole. *Path of the Pole* discusses the various "pole shifts" in Earth's past, giving evidence for each one, and moves on to possible future pole shifts.

356 pages. 6x9 Paperback. Illustrated. $16.95. code: POP

THE CRYSTAL SKULLS
Astonishing Portals to Man's Past
by David Hatcher Childress and Stephen S. Mehler

Childress introduces the technology and lore of crystals, and then plunges into the turbulent times of the Mexican Revolution form the backdrop for the rollicking adventures of Ambrose Bierce, the renowned journalist who went missing in the jungles in 1913, and F.A. Mitchell-Hedges, the notorious adventurer who emerged from the jungles with the most famous of the crystal skulls. Mehler shares his extensive knowledge of and experience with crystal skulls. Having been involved in the field since the 1980s, he has personally examined many of the most influential skulls, and has worked with the leaders in crystal skull research, including the inimitable Nick Nocerino, who developed a meticulous methodology for the purpose of examining the skulls.

294 pages. 6x9 Paperback. Illustrated. Bibliography. $18.95. Code: CRSK

SECRETS OF THE MYSTERIOUS VALLEY
by Christopher O'Brien

No other region in North America features the variety and intensity of unusual phenomena found in the world's largest alpine valley, the San Luis Valley of Colorado and New Mexico. Since 1989, Christopher O'Brien has documented thousands of high-strange accounts that report UFOs, ghosts, crypto-creatures, cattle mutilations, skinwalkers and sorcerers, along with portal areas, secret underground bases and covert military activity. This mysterious region at the top of North America has a higher incidence of UFO reports than any other area of the continent and is the publicized birthplace of the "cattle mutilation" mystery. Hundreds of animals have been found strangely slain during waves of anomalous aerial craft sightings. Is the government directly involved? Are there underground bases here? Does the military fly exotic aerial craft in this valley that are radar-invisible below 18,000 feet?

460 pages. 6x9 Paperback. Illustrated. Bibliography. $19.95. Code: SOMV

REICH OF THE BLACK SUN

Nazi Secret Weapons & the Cold War Allied Legend

by Joseph P. Farrell

Why were the Allies worried about an atom bomb attack by the Germans in 1944? Why did the Soviets threaten to use poison gas against the Germans? Why did Hitler in 1945 insist that holding Prague could win the war for the Third Reich? Why did US General George Patton's Third Army race for the Skoda works at Pilsen in Czechoslovakia instead of Berlin? Why did the US Army not test the uranium atom bomb it dropped on Hiroshima? Why did the Luftwaffe fly a non-stop round trip mission to within twenty miles of New York City in 1944? *Reich of the Black Sun* takes the reader on a scientific-historical journey in order to answer these questions. Arguing that Nazi Germany actually won the race for the atom bomb in late 1944,

352 PAGES. 6X9 PAPERBACK. ILLUSTRATED. BIBLIOGRAPHY. $16.95. CODE: ROBS

THE GIZA DEATH STAR

The Paleophysics of the Great Pyramid & the Military Complex at Giza

by Joseph P. Farrell

Was the Giza complex part of a military installation over 10,000 years ago? Chapters include: An Archaeology of Mass Destruction, Thoth and Theories; The Machine Hypothesis; Pythagoras, Plato, Planck, and the Pyramid; The Weapon Hypothesis; Encoded Harmonics of the Planck Units in the Great Pyramid; High Fregquency Direct Current "Impulse" Technology; The Grand Gallery and its Crystals: Gravito-acoustic Resonators; The Other Two Large Pyramids; the "Causeways," and the "Temples"; A Phase Conjugate Howitzer; Evidence of the Use of Weapons of Mass Destruction in Ancient Times; more.

290 PAGES. 6X9 PAPERBACK. ILLUSTRATED. $16.95. CODE: GDS

THE GIZA DEATH STAR DEPLOYED

The Physics & Engineering of the Great Pyramid

by Joseph P. Farrell

Farrell expands on his thesis that the Great Pyramid was a maser, designed as a weapon and eventually deployed—with disastrous results to the solar system. Includes: Exploding Planets: A Brief History of the Exoteric and Esoteric Investigations of the Great Pyramid; No Machines, Please!; The Stargate Conspiracy; The Scalar Weapons; Message or Machine?; A Tesla Analysis of the Putative Physics and Engineering of the Giza Death Star; Cohering the Zero Point, Vacuum Energy, Flux: Feedback Loops and Tetrahedral Physics; and more.

290 PAGES. 6X9 PAPERBACK. ILLUSTRATED. $16.95. CODE: GDSD

THE GIZA DEATH STAR DESTROYED

The Ancient War For Future Science

by Joseph P. Farrell

Farrell moves on to events of the final days of the Giza Death Star and its awesome power. These final events, eventually leading up to the destruction of this giant machine, are dissected one by one, leading us to the eventual abandonment of the Giza Military Complex—an event that hurled civilization back into the Stone Age. Chapters include: The Mars-Earth Connection; The Lost "Root Races" and the Moral Reasons for the Flood; The Destruction of Krypton: The Electrodynamic Solar System, Exploding Planets and Ancient Wars; Turning the Stream of the Flood: the Origin of Secret Societies and Esoteric Traditions; The Quest to Recover Ancient Mega-Technology; Non-Equilibrium Paleophysics; Monatomic Paleophysics; Frequencies, Vortices and Mass Particles; "Acoustic" Intensity of Fields; The Pyramid of Crystals; tons more.

292 pages. 6x9 paperback. Illustrated. $16.95. Code: GDES

SECRETS OF THE UNIFIED FIELD
The Philadelphia Experiment, the Nazi Bell, and the Discarded Theory
by Joseph P. Farrell

Farrell examines the discarded Unified Field Theory. American and German wartime scientists determined that, while the theory was incomplete, it could nevertheless be engineered. Chapters include: The Meanings of "Torsion"; The Mistake in Unified Field Theories and Their Discarding by Contemporary Physics; Three Routes to the Doomsday Weapon: Quantum Potential, Torsion, and Vortices; Tesla's Meeting with FDR; Arnold Sommerfeld and Electromagnetic Radar Stealth; Electromagnetic Phase Conjugations, Phase Conjugate Mirrors, and Templates; The Unified Field Theory, the Torsion Tensor, and Igor Witkowski's Idea of the Plasma Focus; tons more.

340 pages. 6x9 Paperback. Illustrated. Bibliography. Index. $18.95. Code: SOUF

NAZI INTERNATIONAL
The Nazi's Postwar Plan to Control Finance, Conflict, Physics and Space
by Joseph P. Farrell

Beginning with prewar corporate partnerships in the USA, including some with the Bush family, he moves on to the surrender of Nazi Germany, and evacuation plans of the Germans. He then covers the vast, and still-little-known recreation of Nazi Germany in South America with help of Juan Peron, I.G. Farben and Martin Bormann. Farrell then covers the development and control of new energy technologies including the Bariloche Fusion Project, Dr. Philo Farnsworth's Plasmator, and the work of Dr. Nikolai Kozyrev. Finally, Farrell discusses the Nazi desire to control space, and examines their connection with NASA.

412 pages. 6x9 Paperback. Illustrated. References. $19.95. Code: NZIN

THE TIME TRAVEL HANDBOOK
edited by David Hatcher Childress

The Time Travel Handbook takes the reader beyond the government experiments and deep into the uncharted territory of early time travellers such as Nikola Tesla and Guglielmo Marconi and their alleged time travel experiments, as well as the Wilson Brothers of EMI and their connection to the Philadelphia Experiment—the U.S. Navy's forays into invisibility, time travel, and teleportation. Childress looks into the claims of time travelling individuals, and investigates the unusual claim that the pyramids on Mars were built in the future and sent back in time. A highly visual, large format book, with patents, photos and schematics. **316 Pages. 7x10 Paperback. Illustrated. $16.95. Code: TTH**

DARK MOON
Apollo and the Whistleblowers
by Mary Bennett and David Percy

Did you know a second craft was going to the Moon at the same time as Apollo 11? Do you know that potentially lethal radiation is prevalent throughout deep space? Do you know there are serious discrepancies in the account of the Apollo 13 'accident'? Did you know that 'live' color TV from the Moon was not actually live at all? Did you know that the Lunar Surface Camera had no viewfinder? Do you know that lighting was used in the Apollo photographs—yet no lighting equipment was taken to the Moon? All these questions, and more, are discussed in great detail by British researchers Bennett and Percy in *Dark Moon,* the definitive book (nearly 600 pages) on the possible faking of the Apollo Moon missions.

568 Pages. 6x9 Paperback. Illustrated. Bibliography. Index. $32.00. Code: DMO

THE FREE-ENERGY DEVICE HANDBOOK
A Compilation of Patents and Reports
by David Hatcher Childress

A large-format compilation of various patents, papers, descriptions and diagrams concerning free-energy devices and systems. *The Free-Energy Device Handbook* is a visual tool for experimenters and researchers into magnetic motors and other "over-unity" devices. With chapters on the Adams Motor, the Hans Coler Generator, cold fusion, superconductors, "N" machines, space-energy generators, Nikola Tesla, T. Townsend Brown, and the latest in free-energy devices. Packed with photos, technical diagrams, patents and fascinating information, this book belongs on every science shelf.

292 PAGES. 8x10 PAPERBACK. ILLUSTRATED. $16.95. CODE: FEH

THE ENERGY GRID
Harmonic 695, The Pulse of the Universe
by Captain Bruce Cathie

This is the breakthrough book that explores the incredible potential of the Energy Grid and the Earth's Unified Field all around us. Cathie's first book, *Harmonic 33*, was published in 1968 when he was a commercial pilot in New Zealand. Since then, Captain Bruce Cathie has been the premier investigator into the amazing potential of the infinite energy that surrounds our planet every microsecond. Cathie investigates the Harmonics of Light and how the Energy Grid is created. In this amazing book are chapters on UFO Propulsion, Nikola Tesla, Unified Equations, the Mysterious Aerials, Pythagoras & the Grid, Nuclear Detonation and the Grid, Maps of the Ancients, an Australian Stonehenge examined, more.

255 PAGES. 6x9 TRADEPAPER. ILLUSTRATED. $15.95. CODE: TEG

THE BRIDGE TO INFINITY
Harmonic 371244
by Captain Bruce Cathie

Cathie has popularized the concept that the earth is crisscrossed by an electromagnetic grid system that can be used for anti-gravity, free energy, levitation and more. The book includes a new analysis of the harmonic nature of reality, acoustic levitation, pyramid power, harmonic receiver towers and UFO propulsion. It concludes that today's scientists have at their command a fantastic store of knowledge with which to advance the welfare of the human race.

204 PAGES. 6x9 TRADEPAPER. ILLUSTRATED. $14.95. CODE: BTF

THE HARMONIC CONQUEST OF SPACE
by Captain Bruce Cathie

Chapters include: Mathematics of the World Grid; the Harmonics of Hiroshima and Nagasaki; Harmonic Transmission and Receiving; the Link Between Human Brain Waves; the Cavity Resonance between the Earth; the Ionosphere and Gravity; Edgar Cayce—the Harmonics of the Subconscious; Stonehenge; the Harmonics of the Moon; the Pyramids of Mars; Nikola Tesla's Electric Car; the Robert Adams Pulsed Electric Motor Generator; Harmonic Clues to the Unified Field; and more. Also included are tables showing the harmonic relations between the earth's magnetic field, the speed of light, and anti-gravity/gravity acceleration at different points on the earth's surface. New chapters in this edition on the giant stone spheres of Costa Rica, Atomic Tests and Volcanic Activity, and a chapter on Ayers Rock analysed with Stone Mountain, Georgia.

248 PAGES. 6x9. PAPERBACK. ILLUSTRATED. BIBLIOGRAPHY. $16.95. CODE: HCS

ORDER FORM

10% Discount When You Order 3 or More Items!

One Adventure Place
P.O. Box 74
Kempton, Illinois 60946
United States of America
Tel.: 815-253-6390 • Fax: 815-253-6300
Email: auphq@frontiernet.net
http://www.adventuresunlimitedpress.com

ORDERING INSTRUCTIONS

- ✓ Remit by USD$ Check, Money Order or Credit Card
- ✓ Visa, Master Card, Discover & AmEx Accepted
- ✓ Paypal Payments Can Be Made To: info@wexclub.com
- ✓ Prices May Change Without Notice
- ✓ 10% Discount for 3 or more Items

SHIPPING CHARGES

United States

- ✓ Postal Book Rate { $4.00 First Item; 50¢ Each Additional Item
- ✓ POSTAL BOOK RATE Cannot Be Tracked!
- ✓ Priority Mail { $5.00 First Item; $2.00 Each Additional Item
- ✓ UPS { $6.00 First Item; $1.50 Each Additional Item

NOTE: UPS Delivery Available to Mainland USA Only

Canada

- ✓ Postal Air Mail { $10.00 First Item; $2.50 Each Additional Item
- ✓ Personal Checks or Bank Drafts MUST BE US$ and Drawn on a US Bank
- ✓ Canadian Postal Money Orders OK
- ✓ Payment MUST BE US$

All Other Countries

- ✓ Sorry, No Surface Delivery!
- ✓ Postal Air Mail { $16.00 First Item; $6.00 Each Additional Item
- ✓ Checks and Money Orders MUST BE US$ and Drawn on a US Bank or branch.
- ✓ Paypal Payments Can Be Made in US$ To: info@wexclub.com

SPECIAL NOTES

- ✓ RETAILERS: Standard Discounts Available
- ✓ BACKORDERS: We Backorder all Out-of-Stock Items Unless Otherwise Requested
- ✓ PRO FORMA INVOICES: Available on Request

ORDER ONLINE AT: www.adventuresunlimitedpress.com

Please check: ☑

☐ This is my first order ☐ I have ordered before

Name

Address

City

State/Province Postal Code

Country

Phone day Evening

Fax Email

Item Code	Item Description	Qty	Total

Please check: ☑

Subtotal ▸

Less Discount-10% for 3 or more items ▸

☐ Postal-Surface

Balance ▸

☐ Postal-Air Mail (Priority in USA)

Illinois Residents 6.25% Sales Tax ▸

Previous Credit ▸

☐ UPS (Mainland USA only)

Shipping ▸

Total (check/MO in USD$ only) ▸

☐ Visa/MasterCard/Discover/American Express

Card Number

Expiration Date

10% Discount When You Order 3 or More Items!